THE LOST WORLDS OF
EARTH AND MARS

WILLIAM R SAUNDERS

Tellwell Talent
www.tellwell.ca

ISBN
978-0-2288-5414-2 (Paperback)

ACKNOWLEDGEMENTS

First and foremost, I would like to acknowledge my colleague, George J. Haas, who made the important connection with the geoglyphs on Mars to the earthly Mesoamerican mythology of the Olmec and the Maya. This framework allowed the unraveling of the puzzling geoglyphic art we were discovering in Cydonia, Mars. As well, being an artist, his reproductions of many pictures enabled the publishing of our co-authored books, The Cydonia Codex; Reflections from Mars, in 2005, and our second book, The Martian Codex; More Reflections from Mars, in 2009. He started The Cydonia Institute in 1991 and the website allows for the discussion of new information posted by members.

Of course there would be no knowledge of the Mesoamerican history without the prodigious work of the many archaeologists who, over the last century have uncovered and deciphered the culture and language of the Maya. The work of people like Sylvanus Morley, Linda Schele, David Freidel, Peter Mathews, William and Michael Coe, David Stuart, and many others brought the past to the present.

Then there is the copious volume of work done by author and historian Zecharia Sitchin whose series of books unveiled the ancient past on the other side of the ocean from Mesoamerica and turned legend and mythology into reality.

I would also like to acknowledge the early Mars investigation work by other independent researchers who, like us, had to endure the skepticism and, at times, the ridicule of mainstream scientists in pursuing their work. It is that determination to learn the truth about Mars that helped inspired my research.

Vincent Dipeitro
Greg Molenaar
Erol Torun
Richard C. Hoagland
Dr. Mark Carlotto

I would also like to thank a few others whose work directly or indirectly helped in previous research or with this preparing this book.

Jim Miller
Will Faust
Dennis Atkinson

Finally, thank you to the U.S. and European Space Agencies for providing us with the wonderful images from Mars.

INTRODUCTION

In May of 2014, a document entitled, *Archaeology, Anthropology and Interstellar Communication was* published. This document, a product of the collaboration between NASA researchers, SETI scientists, archaeologists and anthropologists, states that rock art should be included in the search for extraterrestrials. The document states that rock art and carvings should be considered as possible attempts at extraterrestrial communication. This makes sense. If you were an extraterrestrial leaving a message, you would likely want your message to last, and rock, being extremely durable, would be an excellent material with which to work. But if you wanted to be sure your message would survive not only the natural elements, but also destruction from other sources, such as other sentient beings, you would not only build it out of rock, but display it somewhere it could not be accessed, such as at the bottom of the ocean, on the Moon or on Mars.

In 1976 just such an artifact was discovered on the planet Mars. A giant edifice of a human-like face, observed in a photo from the Mars Viking I orbiting spacecraft, became known as 'The Face on Mars'. NASA, however, quickly tried to dismiss it as, "a trick of light and shadow".

In 1998, with the Mars Global Surveyor (MGS) beginning to send back new images of the Cydonia area taken with the Mars Orbital Camera (MOC), my colleague, George J. Haas, and I began sharing our findings which consisted of observations of rock art, not carvings per se, but mosaic-like geoglyphs. Geoglyphs are intelligently created designs or symbols made on the Earth's surface composed from rock, stones and / or soil. They can be found on nearly every continent on Earth, with perhaps the most famous being the Nazca lines of Peru and possibly the most numerous found in the Atacama Desert in Chile.

As well as stretching our imaginations, these Martian works of art were challenging our belief systems. It was becoming hard to confidently classify these structures as geoglyphs because they were not simple geometric designs made by a primitive civilization expressing their day to day lives, but massive structures engineered by a highly advanced culture, more advanced than humans of today. Not only that, but they were telling an elaborate and complex story. The most confounding aspect of this unfolding chronicle was that these colossal works depicted the religious and mythological elements of a civilization that once existed on planet Earth. The civilization, or more precisely, civilizations, were those of Mesoamerica, the Olmec, Maya and Aztec civilizations in particular.

The bloodline of the Aztec and Maya people still exists, but their great civilizations are gone, and though they certainly were great civilizations, I am quite sure they did not have the means for space travel and the engineering of gigantic edifices.

Not only are ancient Mesoamerican tales being told on Mars, but they are presented in the same motif that was, and to some extent, still is used by Central American artisans to create paintings and sculpture. The motif consists of half and bifurcated images along with facial profiles. A combination of the mosaic-like form, the motif and the subject matter make these images hard to see. To make the situation more confusing, some of the glyphic representations tell a Mesoamerican story, but include symbols from other cultures and religions such as the Sumerian, the Egyptian, the Hebrew, and the Hindu. The time frames are also confusing, as some of the symbols are of a more recent period than others.

More recently came the finding of the massive art displays on Earth. Many major discoveries throughout history have occurred by a chance, this was no different. My stumbling across the image of Kukulkan (Quetzalcoatl) was just such a chance occurrence. It led to others, including the magnificent and humbling image on the cover of this book.

I must admit I do not have a clear understanding of the who, the why, the when, and the how of these incredible displays. My hope is that others who read this book may be able to offer a deeper understanding and address some of the questions that arise herein. What I do know, and what you hopefully will discover from reading this book, is that neither science nor religion, tell a true story of our history.

William R. Saunders

"Science is about recognizing patterns. [...] Everything depends on the ground rules of the observer: if someone refuses to look at obvious patterns because they consider a pattern should not be there, then they will see nothing but the reflection of their own prejudices."

— Christopher Knight

"Mirrors symbolize reality, the sun, the earth, and its four corners, its surface, its depth, and all of its peoples. Buried in caches throughout the Americas, they also cling to the bodies of the humblest celebrators in the Peruvian highlands or in the Mexican Indian carnivals. ... Are they not right? Is not the mirror both a reflection of reality and a project of the imagination?"

— Carlos Fuentes, *The Buried Mirror*

TABLE OF CONTENTS

CHAPTER I

THE FACE THAT LAUNCHED A THOUSAND QUESTIONS

Did you see, did you see, the face on Mars
Lookin' out into space beyond the stars
Did you see, did you see, the face on Mars
We could be anyplace beyond the stars
– Cody Canada and the Departed

THE 1976 FACE ON MARS

According to both the Ancient Greeks and the Ancient Maya, we are currently living in the 5th age of man. Each age was destroyed by cataclysms of one type or another. Both cultures insisted that there was a time, long ago, when man and the gods lived together on Earth in harmony; this was the first age, the Golden Age.

We had to go to the planet Mars to actually learn that what we think we know about humankind is wrong. Neither our science nor our religions tell us the whole truth about who we are and from where we come. This book will not necessarily answer those questions, but it will prove that many of the mythologies of our ancient cultures had a basis in fact. The ancient texts were recording history, not fairy tales. Granted, it is unwise to take all of the stories literally, but the tales can be understood as either factual or based on real events.

A well-known line from a 1604 play, *The Tragical History of the Life and Death of Dr. Faustus* by Christopher Marlowe asks, *"Was this the face that launched a thousand ships?"* This famous line refers to the face of the beautiful Helen of Troy. She was the Queen of Sparta, who was abducted

1

by Prince Paris of Troy after she was promised to him by the goddess Aphrodite. It is said that Sparta launched a thousand ships to sail to Troy to free her. Historians generally considered this story, from Homer's Iliad, to be nothing more than romantic mythology, for after all there was no city of Troy and there certainly could not have been a goddess Aphrodite! However, that line of thinking began to change when, in 1878, German archaeologist Heinrich Schliemann began excavations in Turkey, excavations that eventually unearthed the ancient city of Troy.

On July 25th, 1976, another face, an unworldly face, would launch a thousand questions. This face was not discovered by archaeologists, but by NASA, on an oddly shaped Martian mesa in an area called Cydonia. The mesa was photographed from an altitude of 1000 miles (1600 km) by the Viking 1 spacecraft, which was orbiting the planet Mars. The photograph revealed an image that looked remarkably like a human face. This face-like mesa, approximately a mile and a half long and a mile wide, was initially spotted on Viking frame 35A72 by a member of NASA's own imaging team at the Jet Propulsion Laboratory (JPL) in Pasadena, California.

While searching for a possible landing site for the upcoming Viking 2 Lander in the Cydonia region, Dr. Tobias Owen noticed a gigantic, human-like face glaring up at him from the barren Martian surface (Figure 1.1). Planetary geologist and head of the Viking Orbiter imaging team, Michael Carr, immediately released the unusual image to the press.

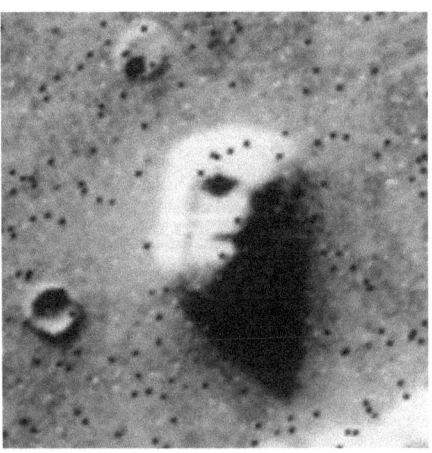

1.1 The 'Face on Mars', 1976 Viking Orbiter, frame 35A72. Courtesy NASA/JPL/CalTech.

2

Soon after its discovery, NASA's Viking project scientist, Gerry Soffen announced at a press conference that an image of an odd landform resembling a face had been found. However, the press and news media were quickly informed that when a second picture was taken only a few hours later, the image of the "Face" had disappeared. Oddly, this second image never surfaced.

NASA's official position was that the "Face" was a trick of light and shadow, and the overall mesa had no resemblance to a human face. What they didn't say was that, like the equinox serpents, on the pyramid of Kukulkan in Chichen Itza, it was an intentionally designed trick of light and shadow. Consequently, only a single high-contrast picture of the "Face" was circulated to the press and it was dismissed as nothing more than a phantom novelty.

Conveniently, NASA decided not to proceed with the original plan of a Viking 2 landing at Cydonia because the area was deemed to be "unsafe." The Viking 2 Lander was eventually set down on a rocky plain called Utopia. This last-minute change in plans went virtually unquestioned by the media. With NASA's firm and consistent stance that the face-like landform was nothing more than an apparition of shadows and rock, the public soon lost interest. However, there were a few independent scientists who were not satisfied with the explanation, and they began to study these fascinating images of our neighboring planet on their own.

Researchers looking at the structures in Cydonia made the assumption that, if there was anything of artificial creation, it had to be very old and highly eroded. They made this assumption because they didn't see anything that looked like a complete structure. This was human bias. They also assumed that if the Face was meant to be a face, it should have been a bisymmetrical, carved structure, much like the U.S. presidents on Mount Rushmore. Once again a bias, only this time cultural. The Face is huge and staring up at the sky, obviously meant to be seen from above, not from the ground. If one is viewing a large image from a distance it does not have to have fine detail, because these details merge together as you move further away. It is like viewing a large mosaic, one needs to be within the proper viewing distance before its scattered pieces begin to form an image the brain recognizes. Image clarity relies on the observer's viewing perspective.

Science journalist, Richard C. Hoagland began his own research into the 'Face on Mars' and the Cydonia area in the 1980's. He also assumed the mesa supported a symmetrical face whose one side was eroded more than the other. Hoagland acquired a photograph of the Face, which was a computer enhancement by imaging specialist Dr. Mark Carlotto. In an effort to create symmetry, he mirrored the left side of the image (Figure 1.2). Impressed by the humanoid face that resulted, he mirrored the opposite side. This side was supposed to have been highly eroded. What he was presented with shocked him. The mirrored image staring back at him was the depiction of a lion (Figure 1.2)! Hoagland concluded that the Face on Mars was reminiscent of the Egyptian Sphinx, a human – lion hybrid. It seemed far too unlikely to be just a coincidence. The right side of the face was not highly eroded, and it was not a human face.

THE 1998 FACE ON MARS

It was another twenty-two years after the Viking Orbiter photograph of 'The Face' before we were able to see it again (Figure 1.3). The Viking image of 1976 was taken from one thousand miles up, whereas the new image was from a distance of 250 miles. This new image, from the Mars Orbital Camera (MOC) onboard the Mars Global Surveyor (MGS) spacecraft, was also of a much higher resolution.

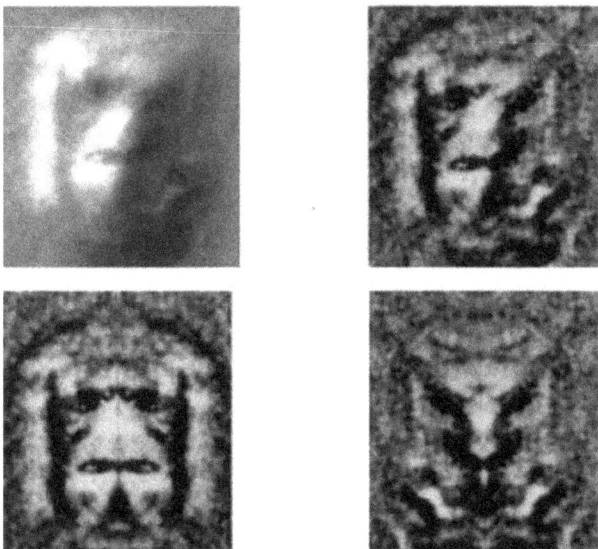

1.2 Top left: 'Face on Mars' (NASA Viking frame 70A13).
Top right: Face on Mars (NASA Viking frame 70A13 contrast
enhanced) Bottom left: mirrored left side: humanoid appearance.
Bottom right: mirrored right side: feline appearance.
Courtesy of Dr. Mark J. Carlotto (TASC) and Richard C. Hoagland.

Immediately upon acquiring access to the new image of the Face, my colleague George Haas and I performed a mirrored split of the image to see if Hoagland's sphinx was still there (Figure 1.4). Not only was it still there, it was even more obvious that we were seeing a humanoid / lion split faced edifice.

Both lions and humans are creatures of Earth, so what could be the connection between this structure and Earth? A few researchers, including astrophysicist Carl Sagan, noticed that some of the other structures in Cydonia and elsewhere on Mars resembled pyramids.[1] Pyramids and a human-lion hybrid would automatically lead most back to ancient Egypt. However, it was the tri-pointed symbol on the forehead of the mirrored humanoid side that George Haas identified as a symbol of royalty used by the ancient Olmec and Maya civilizations of Mesoamerica! The triadic, leaf-shaped glyph denotes the sprouting maize seed and the transformational properties of corn. [2]

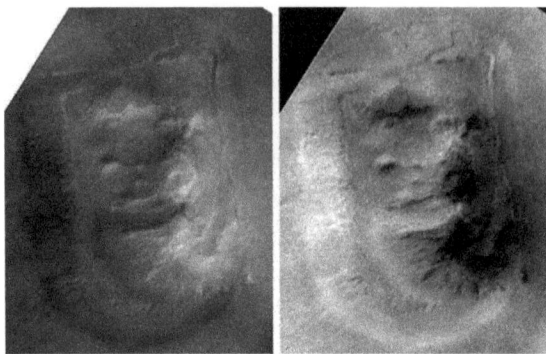

1.3 NASA's enhanced version of the 1998
"Face" from the MOC (SP-22003).
Left: NASA's JPL-enhanced image. Right: contrast-reversal image.
(Courtesy of NASA/JPL/Caltech).

As seen in a Greenstone mask of the 1ˢᵗ century B.C.E. (Before Current Era), the Maya exhibited a three-pointed leaf emblem on their headbands which signified the crown of early Kings (Figure 1.5). Haas further learned that split masks were an artistic motif used in Mesoamerican art, religion and mythology. There were abundant examples, but one that especially caught his attention was a split faced human – feline mask not unlike the Face on Mars. There are also pieces of artwork that are intentionally only half of an image, including their glyph for a mirror (Figure 1.6). This motif is called Herencia, which means cut-in-half.

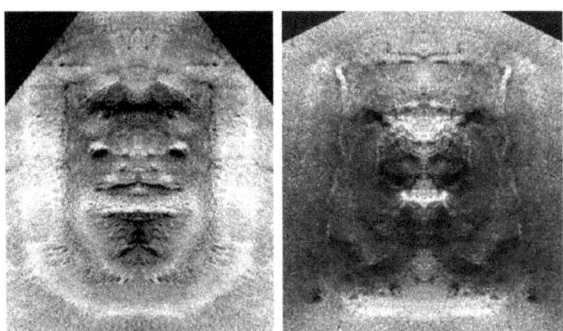

1.4 Left side of the "Face" mirrored. Humanoid face–contrast reversal.
Right side of the "Face" mirrored. Feline face–contrast reversal.
(*The image of the "Face" used to create this split image is an
enhancement of the JPL version courtesy of Richard C. Hoagland*).

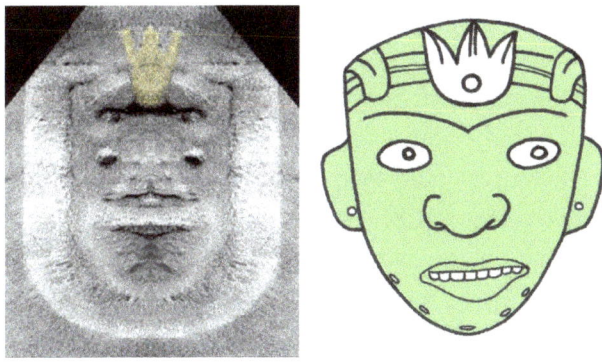

1.5 Left: mirrored humanoid face on Mars with tri-leaf emblem.
Right: Maya greenstone mask with tri-leaf symbol of royalty.

1.6 Left: Pre-Columbian split-faced human-feline mask,
International Museum of Ceramics, Florence, Italy.
Center: Olmec half-faced jade pendant, drawing by George J. Haas,
(Image source, Olmec and Their Neighbors, Coe and Grove, 319.)
Right: mirror glyph, represented by half a
mirror, drawing by George, J. Haas,
(Image source: Blood of Kings, Schele & Miller, p 283).

Some researchers have been concerned about the absence of a distinct nose formation on the Face. They have speculated that the nose was blown off sometime in the past by a meteorite or perhaps by the act of some ancient Martian war. The debris or fallout of this major hit distorted the nose and left an odd feature, which Hoagland called the "tear", resting on the cheek.[3] However, could this remnant actually be part of an intentional design?

Haas and I believe what we are seeing is a large ceremonial nose ornament that obscures the nose. The tear, or teardrop, is just one part of a larger facial ornament covering the entire nose area. This type of ornamentation over the nose is typical of the ones used by the Maya, Aztec (Figure 1.7) and most notably by the Tiarona Indians of Columbia's Santa Marta mountains (Figure 1.8).

The lion has been associated with the Sun and Sun gods since ancient times. It is easy to see its round face as a disk and its mane as the radiant flare. In Syria the lion was the symbol of the Sun and the Semitic god of the Sun, Shamash, could be portrayed as a lion.[4] The Egyptian Sun god Ra, was associated with the lion, and the lioness was the female aspect of the Eye of Ra.

After revealing two cultural ties to the Americas, within the design of the Face on Mars, how does one account for a lion? Perhaps it represents another large cat, a jaguar? In the Maya culture, a Bearded Jaguar was seen as a Sun god. Not only does it seem an unlikely coincidence that the Sun should be associated with a large cat in two distant cultures, but how do we account for the existence of a bearded jaguar? Jaguars don't have beards so where did this idea of a bearded jaguar come from? It is my opinion that this idea came from an outside source telling them about the jaguar's cousin in another land, and therefore the Bearded Jaguar = Male African lion = The Sun.

1.7 Butterfly–Aztec gold nose ornament.
Note: The rod went through the septum of
the nose to support the adornment.
Drawing by George J. Haas.
(Image source: *Aztecs: Reign of Blood & Splendor*,
editors of Time-Life, page 138).

1.8 Tairona warrior's gold pendant.
Notice the segmented nose ornament and oval chin adornment.
Drawing by George J. Haas. (Image source: *Lost Empires,
Living Tribes*, National Geographic Society, page 176).

1.9 Former Emblem of Iran
By MrInfo2012 - Own work, CC BY-SA 4.0, T
https://commons.wikimedia.org/w/index.php?curid=57028181

1.10 Male African lion; bearded jaguar?

In the National Museum of Anthropology in Mexico City sits a large Aztec reliquary that is carved in the shape of a lion. However, because of its geographical and cultural setting, it is officially labeled a jaguar (Figure 1.11). This impressive sculpture, which weighs over six tons, was unearthed at Temple Mayor in Mexico City in 1790. The most intriguing characteristics of the so—called jaguar sculpture are that it has no spots, as jaguars do, and it has a mane, which jaguars do not. [5]

1.11 Aztec jaguar reliquary, side and front views. Note the absence of jaguar spots and the presence of a mane. Drawing by George J. Haas, (image source: *Myths of The World; Gods of the Inca, Aztec and Maya*, by Roberts, p 67).

THE FIRST LORD AND JAGUAR SUN

On a two-tiered ancient Maya pyramid that was intentionally buried around 50 B.C.E., are some fascinating sculptures. Archaeologists call this pyramid the First Temple (Figure 1.12). It is located on a small peninsula at Cerros in what is now Belize. Occasionally the Maya practiced a strange ritual of urban renewal in which they would actually bury temples or entire villages, and then build new structures over them. Because of this practice, the elaborately carved figures on the Temple are in relatively good shape.

1.12 Left: First Temple at Cerros unexcavated, photo courtesy David Schele; Right: First Temple at Cerros, excavated, photo courtesy Megan L. Wood.

On the face of the Temple, four decorated panels with large plaster-covered masks flank a central stairway (Figure 1.12). The top two masks represent the planet Venus: as the Morning Star on the eastern side and the Evening star on the western side. The lower masks represent the Jaguar Sun god: as the rising sun in the East and the setting sun in the West. Venus and the Jaguar Sun are also representations of the Maya twin gods known as First Lord (Venus) and First Jaguar (Jaguar Sun). [6] First Lord was the God associated with resurrection while First Jaguar was the god associated with the Underworld and death. This temple-pyramid is seen as a cosmological diagram that joins the heavens with the underworld.

In describing the First Temple at Cerros in their book, *A Forest of Kings: The Untold Story of The Ancient Maya*, mayanist Linda Shelley and archaeologist David Freidel state:

> "*The First Temple was in the center of the vertical axis that penetrated the earth and pierced the sky, linking the supernatural and the natural world into a whole. This plan set the Temple between the land and the sea on the horizontal axis and between the heavens and the underworld on the vertical axis.*"[7]

The panels were designed to be read as symbolic statements about the nature of the kingship and its relationship to the cosmos.

Within the design of the First Temple at Cerros, with its symbolic pairing of the Sun and Venus as the Cosmic Twins, we discover the same fusion of human and feline aspects that we find in the Face on Mars.

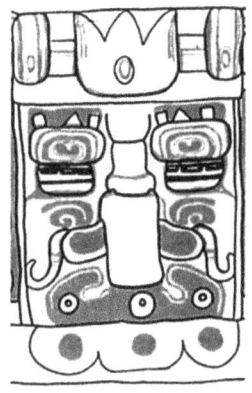

1.13 Left: Temple mask: Morningstar (First Lord / Venus). Notice the triad crown emblem and the odd facial ornaments around the nose, mouth and chin area. Drawing by Linda Schele, © David Schele. (Courtesy of Foundation for the Advancement of Mesoamerican Studies, Inc., 2002.) Right: Temple mask: Jaguar Sun. Note the snarling aspect of the snout and the glyph tag under the neck. Drawing by Linda Schele, © David Schele. (Courtesy of Foundation for the Advancement of Mesoamerican Studies, Inc., 2002).

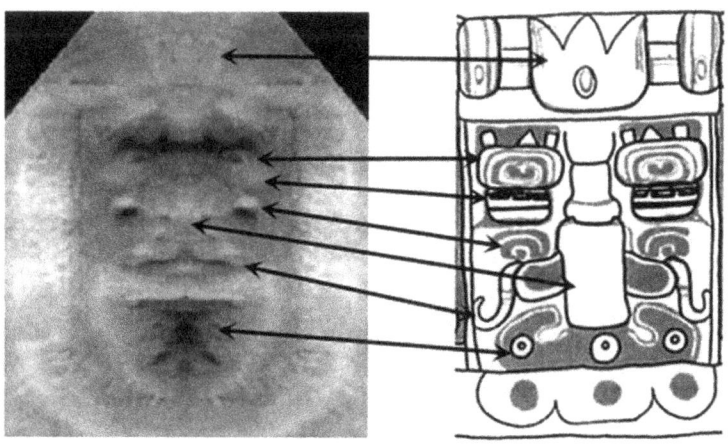

1.14 Comparison to mirrored humanoid side of the 'Face on Mars' and First Lord / Venus Cerros Temple Mask.

As you can see, there are numerous points of commonality between the Morning Star mask on the Maya Temple and the mirrored humanoid Face on Mars. Both present the same triad crown emblem on the headband, the depiction of the staring eyes, the bulge or puffiness under the eyes and similar facial ornaments in the nose, mouth and chin area, including the teardrop feature on the cheek (Figure 14).

When the completed feline side of the Face is compared to the Jaguar Sun mask on the lower panel of the Temple, it is apparent that, again, many points of commonality are displayed (Figure 1.15).

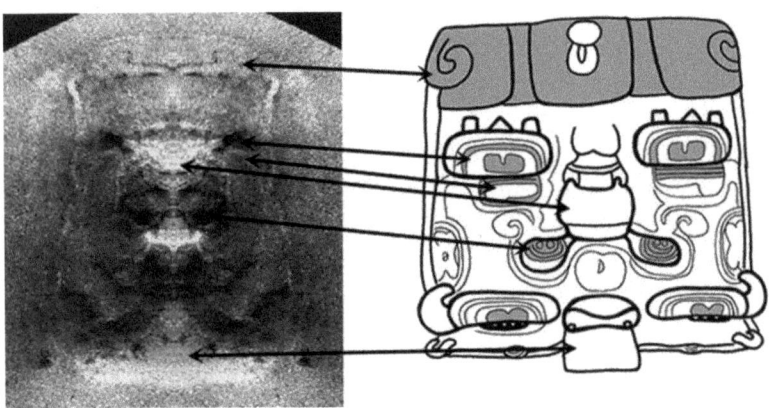

1.15 Comparison to mirrored feline side of the Face on Mars and Jaguar Sun temple mask.

The Jaguar Sun god is portrayed as both the Bearded Jaguar and the Sun. If you look at just the circular muzzle and the flaring face of the Mars feline, a comparison to the common diagrammatic representation of the sun is easily made (Figure 1.16). As you will see in a few other examples, shape plays a role in the symbolism presented in these Martian glyphs.

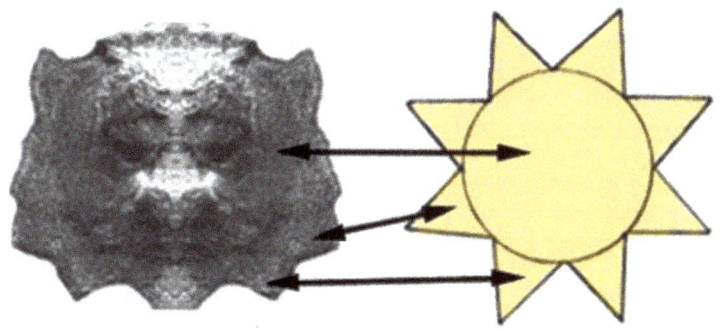

1.16 Comparison of the mirrored feline muzzle on the 'Face on Mars' to a common diagrammatic depiction of the Sun with its central circle and flaring points.

On January 31, 2001 Malin/NASA/JPL released seven additional Cydonia images; these had been taken after the images released in April 2000. Among this latest deluge of Cydonia images was a fine-detail photograph of the western (humanoid) side of the Face, labeled M16–00184 (Figure 1.17). This was the highest-resolution picture ever taken of this controversial structure, and Haas and I feverishly began to analyze it.

Although disappointing in its lack of totality, this image presents details at 1.7 meters or 5.6 feet per pixel and is quite revealing.

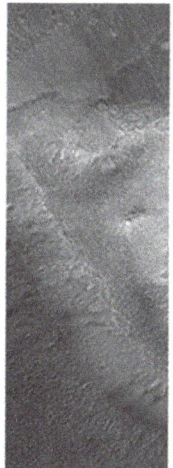

1.17 M16-00184 Courtesy NASA/JPL/Caltech.
(Contrast reversal by author).

The new image confirms the existence of the headdress, the triad-leaf crown emblem, the "teardrop" feature, and an eye on the humanoid side of the Face. The image also captures a tiny corner of the mouth.

In the forehead area, notice the oval "gemstone" marking a portion of the half emblem that forms the triad-leaf symbol. When mirrored, this half emblem completes a "W"-shaped, triad-leaf crown emblem in which the "gemstone" acts as ornamentation.

The most positive result of this new image is that it captures the brow area, complete with confirmation of an almond-shaped eye. The central bulge forms a pupil, while the surrounding depression creates an iris (Figure 1.18).

Another pleasant surprise on the M16 image is a "trophy head" mounted just above the eye. The deer bust forms the brow area of the humanoid side of the Face. An analytical drawing of these intriguing facial features is provided in figure 1.19.

The symbolic significance of the deer effigy placed above the "eye" is not completely clear at this point, but the deer was a sacred symbol to the Maya and was sometimes associated with death and re-birth, but also, like the tri-leafed emblem, was associated with royalty and power.[8] Also of interest, throughout the Middle East, stags and gazelles were used as emblems to denote sharp eyesight.[9]

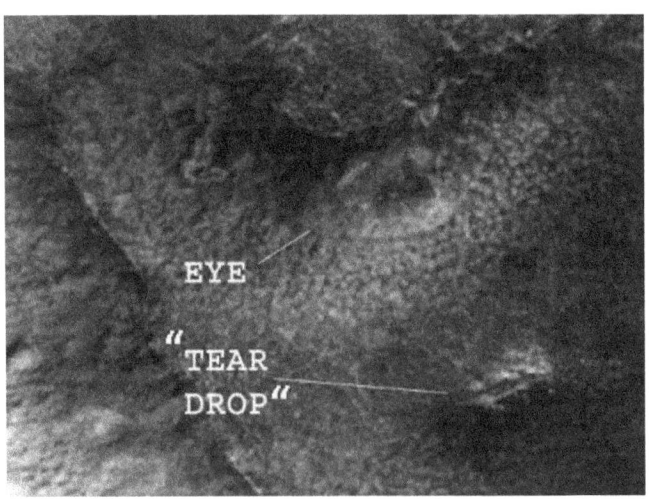

1.18 Detail of the eye and "teardrop" feature.
Cropped from M16-00184.

1.19 Detail of the eye and "teardrop" feature.
Drawing by George J. Haas.

THE 2001 FACE ON MARS

A new, complete image of the Face was present in another image released in 2001, MOC E03-00824 (Figure 1.20). This image is from nearly directly above and gives further confirmation of its bifurcated structure.

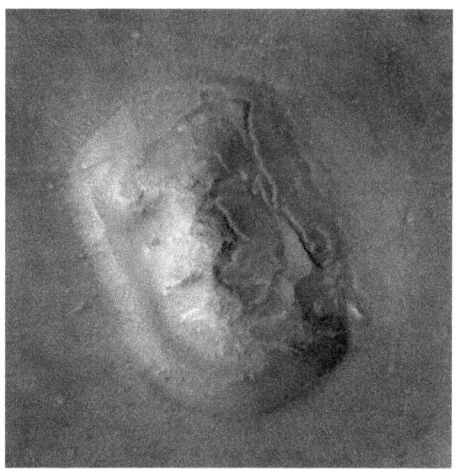

1.20 2001 'Face on Mars'.
(E03-00824 courtesy NASA/JPL/Caltech).
Enhancement courtesy of Keith Laney.

As soon the new image hit the web, every attempt to mirror the new 2001 Face image was cropped either too wide or too narrow. Many advocates of a bifurcated Face who mirrored the feline side of the geoglyph totally disregarded any sense of a central axis by including portions of the triad-leaf symbol (the "W") from the "humanoid" side.

If you look closely at the forehead of the humanoid side of the Face, there is an oval, appearing like a gemstone, located within the half emblem design of the triad-leaf symbol. As demonstrated earlier, this half emblem completes a "W"-shaped, triad-leaf crown emblem when mirrored. The "gemstone" is part of the emblem's ornamentation and is not a demarcation line marker.

It has been our experience that all of these bifurcated geoglyphs have precisely placed demarcation lines that are signaled by markers, such as mounds and/or grooves, that establish the line of symmetry. We agree with Hoagland that there is a centerline; however, the oval "gemstone" within the forehead area is not such a marker.

The Face has three notable markers that establish a demarcation line. The first runs along the edge of the half emblem of the triad-leaf symbol ("W") identified at the center of the forehead on the "humanoid" side of the Face. A second marker sits at the edge of the central tooth feature on the humanoid side and the protruding tongue and fang on the feline side. The last is a small vertical bar located at the edge of the mane, placed between the "feline" and "humanoid" sides of the Face.

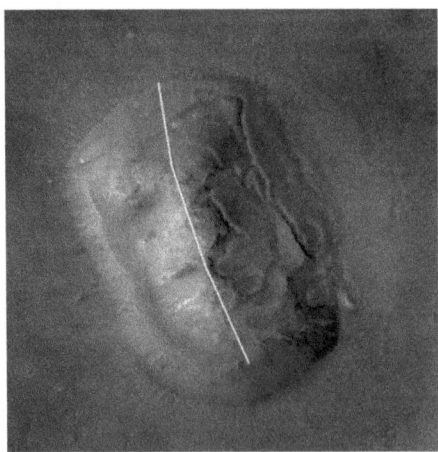

1.21 The line of demarcation of the split-faced mask.

You will find this marker, which appears as two vertical parallel lines, in many of the incorrectly mirrored splits of the feline side of the "Face." This vertical bar denotes the central axis (or marker) between the two half faces. It should also be noted that because of the three-dimensional aspect of this bifurcated facial structure and the camera angle, the demarcation line would appear irregular as it arches across this split-faced edifice (Figure 1.21).

FOOTNOTES

1. In his 1980 book *Cosmos*, Carl Sagan shows a picture in the chapter *Blues for a Red Planet* with the image caption "The Pyramids of Elysium". In a footnoote on that page he wrote: *The largest Mars Pyramids have a base width of 3km and a height of 1km, so they are much larger than the Pyramids of Sumer, Egypt and Mexico. With the ancient eroded shape, they could be small hills, sandblasted for centuries, but they need to be viewed from nearby.*
2. Linda Schele and David Freidel, *A Forest of Kings: The Untold Story of the Ancient Maya* (New York: Quill, 1990), 115.
3. Richard C. Hoagland, *The Monuments of Mars: A City on the Edge of Forever*, 5th ed. (Berkeley: North Atlantic, 1992), 22.
4. Babayan, Kathryn, *Mystics, Monarchs, and Messiahs: Cultural Landscapes of Early Modern Iran*, (Harvard College, 2002) 49
5. All jaguars have spots except for the Black Jaguar.
6. Linda Schele and David Freidel, *A Forest of Kings: The Untold Story of the Ancient Maya* (New York: Quill, 1990), 117
7. Ibid, 105
8. Deer ephigy: https://mayansandtikal.com/mayan-society/mayan-animals-animals/
9. Jean Chevalier and Alain Gheerbrant, *A Dictionary of Symbols* (New York: Penguin Books, 1996), 923. The stag and the gazelle are often regarded as animals with common symbolic attributes, keen eyesight being just one of them. Also see the Bible, Song of Solomon 2:9. In nature, the stag is the victim of the lion, who is its predator. Notice that a feline face lies on the eastern side of this bifurcated geoglyph.

CHAPTER II

THE FOURTH CREATION

The original book, written long ago, existed but its sight is hidden from the searcher and from the thinker.[1]
       - Popol Vuh, *The Ki che' Maya Bible,*

"Mirrors symbolize reality, the sun, the earth, and its four corners, its surface, its depth, and all of its peoples. Buried in caches throughout the Americas, they also cling to the bodies of the humblest celebrators in the Peruvian highlands or in the Mexican Indian carnivals... Are they not right? Is not the mirror both a reflection of reality and a project of the imagination?"
      — Carlos Fuentes, *The Buried Mirror*

Steganography is a word which has become known in more recent years referring to generally illicit computer files being hidden within other files. The word is derived from the Greek words *steganos* (meaning *hidden* or *covered*) and the Greek root *graph* (meaning *to write*). So this would translate to hidden writing. It's meaning, however covers anything which is <u>hidden in plain sight.</u>

In our first two books, The Cydonia Codex and The Martian Codex, George J. Haas and I presented dozens of examples of art work that are hidden in plain sight on the surface of Cydonia, Mars. The creators of this stunning art work presented it in the same motif of bifurcated, half and profiled images used by the ancient Mesoamericans and even still today by artisans of Mexico and Central America. The Mars art work, like the Maya writing system, is very complex and even after over 20 years of research we are still uncovering new images in areas previously searched. The art work is laid out in ways that we are not accustomed. To be fully appreciated imaging software is needed for the adjustment of contrast and brightness, rotating images, and enabling the cropping, duplicating and reconstructing. i.e., mirroring. Many of the images overlap upon others and contour rivalry* is common place. It is mind boggling as to how all this could have been laid out and created.

THE LID OF PACAL

Early on in our investigation of the surface of Cydonia, we came across a book entitled *The Mayan Prophesies* by Maurice M. Cotterell. In 1992 he declared that he had broken the code of the famous Lid of Pacal (Figure 2.1). This fantastic sarcophagus was originally discovered in 1952 at Palenque by freelance Mexican archaeologist Alberto Ruz.

In recent years, the carved Lid of Pacal has become well-known because of the speculations of its connection with ancient space flight. Many believe the Lid depicts the Maya king, Pacal, working a control panel while sitting in what appears to be a flying vehicle. Skeptics say that this interpretation is pure fantasy. Most scholars contend that the Lid depicts the king at "the moment of death as he falls from the world of living to the Underworld."[2]

In Cotterell's unorthodox analysis of the inscriptions, he proclaimed that concealed within the complexity of the Lid's design is an inscribed matrix of the original Popol Vuh of the Maya. The Popol Vuh, considered to be the "Maya Bible," was written in Mayan hieroglyphics on accordion-folded books that documented the history of the Quiche (Ki che) Maya. Portions of the original text and those that had been transcribed into Spanish in the sixteenth century were saved from destruction when the Spanish burned many of the Maya records during their conquest. According to text in the Preamble of the Popol Vuh, . . . The Popol Vuh cannot be seen anymore…The original book, written long ago, existed, but is now hidden from the searcher and from the thinker. . . .

---

* **Contour rivalry** is an artistic technique used to create multiple possible visual interpretations of an image. An image may be viewed as depicting one thing when viewed in a certain way; but if the image is flipped or turned, the same lines that formed the previous image now make up an entirely new design.

Cotterell interpreted this cryptic sentence to mean that perhaps the "hidden" images he was finding on the Lid of Pacal were actually the concealed pages from this hidden book. He discovered that through the use of transparent overlays on the Lid of Pacal, images of Maya gods could be found that paralleled the creation story of the Popol Vuh.

He demonstrated that when the "broken corners" of the Lid of Pacal are overlapped, with the aid of transparencies, the meaningless half figures and partial glyphs along its edge burst into life.[3] When Cotterell performed an overlay of the top border boxes of the Lid, a totemic set of faces emerged, including a bird and human head, followed by a snarling tiger and a dog head. The four faces were created out of three portraits of Maya lords. According to Mayanist Linda Schele, these three individuals have identities: they are a court official and two administrators, the latter two each identified as a "Keeper of the Holy Books."[4] In this context, where the Lid of Pacal is perceived as a multifarious matrix of a sacred book, the identity of these two lords as keepers of "Holy Books" extends tremendous support to Cotterell's belief that hidden images of the Popol Vuh are indeed incorporated within the design of this remarkable lid.

Cotterell found that the same results were achieved when the main body of the Lid was overlaid at different key points. It was not until these secret glyphs were matched up with their second halves (like a mirrored image) that it was possible to see what the glyphs were meant to represent. After a substantial number of the "hidden" images had been brought to light and identified by Cotterell, it became clear to him that the "breaking" of the corners was a deliberate and intentional part of the Lid's design.

One of the missing corners of the Lid had the remains of a broken glyph that Cotterell called a "defect" marker. This marker was part of a crossed pattern of dots that would form a complete "X" if it was not broken. When Cotterell joined the corners of this marker through the use of acetate overlays, he not only restored the "defective" pattern, but the three Maya lords and the partial glyphs along the border boxes transformed into the group of stacked faces. Then in the inner portion of the carving he detected another "defect" mark on the ridge of the central characters nose. After placing the acetate drawing over the "defect" mark, along the nose, the image of the Maya Bat God remarkably appeared (Figure 2.2).

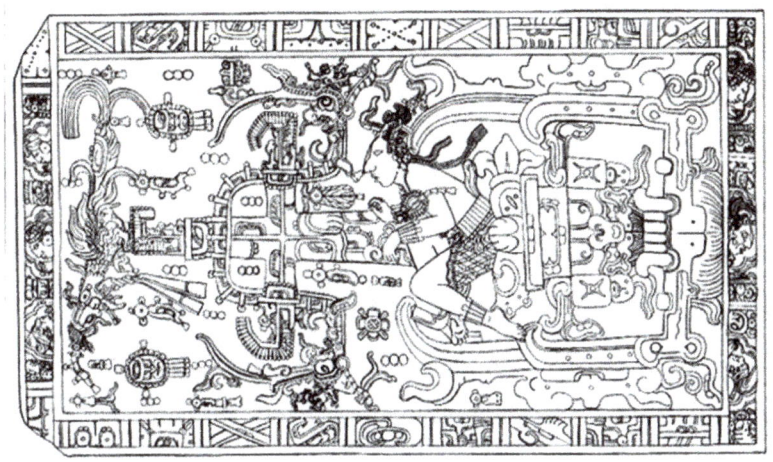

2.1 Lid of Pacal.
Note the "broken" corners of the lid (upper right and left).
Drawing by George J. Haas.
(Images source: photograph by Merle Green Robertson).

2.2 Maya Bat god; Lid of Pacal overlay.
Overlay and coloring by Jim Miller (after Cotterell).

In coming across this work of Cottrell's, not only did we learn that someone other than ourselves had discovered images of Maya religion and mythology that had been hidden, but the use of transparent overlays was remarkably similar to our use of mirroring half and bifurcated images. There was another major similarity, as you will see as we progress in this book, and that is the formation of an 'X' when an image is completed, often occurs.

Further confirmation that we were not alone in our methodology was this quote by Mesoamerican archaeologist Francisco Valdez in referring to an unusual bifurcated design on a Vase found in an archaeological site in Ecuador dating back to 3500 B.C.E. Referring to the motif as "bipartite division" he states:

> *"This symmetry is not only opposition, but a form of complementarity expressed in a very particular way. If an image is spread out symmetrically its whole personality can be revealed. Using the technique of mirror projection it is possible to complete the outline of a figure by presenting its image in reverse."* [5]

To our surprise, even though we were uncovering some stunning imagery, our work was being criticized and rejected by other researchers. We were told, "It is simulacrum, Rorschach imagery, pareidolia" and other convenient terms to dismiss our work without even investigating. However, we knew we were right and the critics were wrong, so the research continued.

THE SQUARE AND COMPASSES

With the new images from the Mars Global Surveyor, besides being able to confirm the human-feline split-faced Face on Mars, perhaps one of the most important images we came across was found in April of 1998 after the release of the Mars Orbital Camera's first images of Cydonia. Some unusual markings were noticed on the surface in an area which had been deemed "The City" by Richard Hoagland (Figure 2.3).

The markings were not recognizable on their own, but on a hunch, we held a mirror up to the photo and low and behold the markings transformed into the depiction of a highly ornamented set of square and compasses. The square and compasses are tools of geometry and architecture but are well-known as the most identifiable symbol of the Freemasons, an organization that has root connections to the Master Builders and ancient Egypt. They are also symbols of opposing elements, the square and the circle. The duality is extended further in that the circle represents the heavens and the square represents the Earth.[6] In regards to compasses, J.E. Cirlot, in *The Dictionary of symbols*, states:

> *"An emblematic representation of the act of creation, found in allegories of geometry, architecture and equity. By its shape it is related to the letter A, signifying the beginning of all things. It also symbolizes the power of measurement, of delimitation."*[7]

2.3 Markings on surface of Cydonia, Mars (MOC SP1-25803).

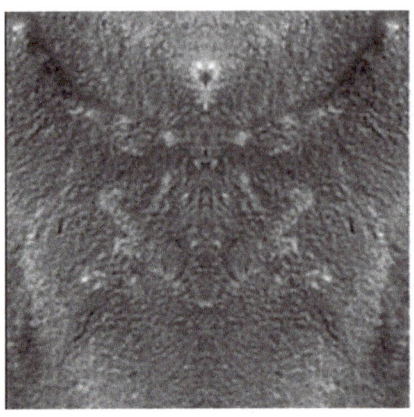 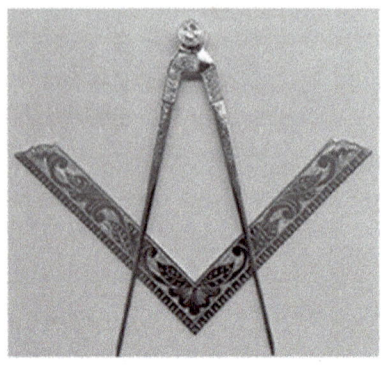

2.4 Left: Mirrored square and compasses on Mars; Notice the artistic rotation effect of the top of the compasses. Right: Square and Compasses (Courtesy, Macoy Publishing and Masonic Supply Co. www.macoy.com)

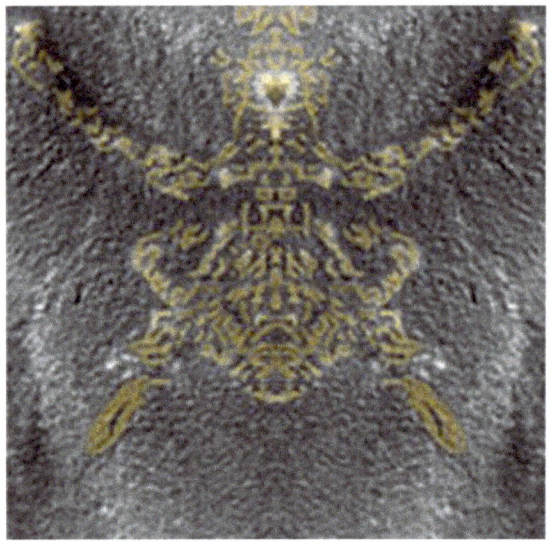

2.5 Mirrored compasses and square on Mars. False color by author.

A Mayan linguist, Domingo Martinez Paredes, maintains that the Maya had a cosmic principle of movement and measurement that symbolized a dualist view of the universe that they called Hunab Ku (Sole Measure Giver). They believed in a dynamic dualism in which the whole material world was part of a cosmic mathematical order. The Hunab Ku was symbolized by a square in a circle (Figure 2.6). According to Martinez, the Hunab Ku and the Masonic square and compasses are synonymous concepts that symbolize the standards of the "Great Architect of the universe."[8]

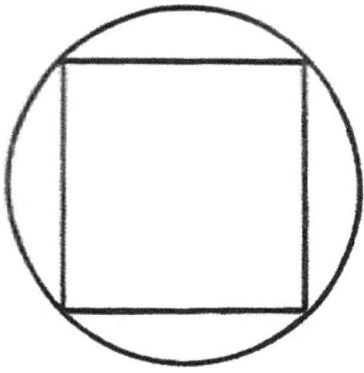

2.6 Maya Hunab Ku (Square in a Circle)

2.7 God as the Great Architect of the Universe (thirteenth century French illuminated manuscript of the Bible, Pierpoint Morgan Library) Note the compasses being used to measure the universe. Drawing by George J. Haas, (Image source: Mythology, by David Leeming, 1976), 152

THE DOLPHIN

When the new MOC images of Cydonia were first released to the public one of the first surprising finds was the figure of a Bottlenose dolphin incised into the Martian surface (Figure 2.8). Two separate, independent researchers, contributing to the Enterprise Mission forum, noticed this dolphin shape. The dolphin is about three kilometers or two miles in length and the image includes a prominent dorsal fin, a flipper, a crescent-shaped tail and a bottle-shaped nose.

The location of the compasses and square markings immediately above the dolphin's head was a bit of a puzzle, however, duality may again play a part. Although dolphins may look like a fish, they actually fall within a class of aquatic mammals. The Olmec and the Maya may have found this transformational aspect of the dolphin appealing because of its tendency to leap out of the water, giving the appearance of living between two worlds where the "sea and the sky meet." This dual identity of the dolphin may have been one of the reasons the Olmec incorporated its shape within their ballcourt iconography. The ball-game's courtyard was seen as a stage for the re-enactment of Creation, and the dolphin may also be seen as a symbol of that transformation and Creation.

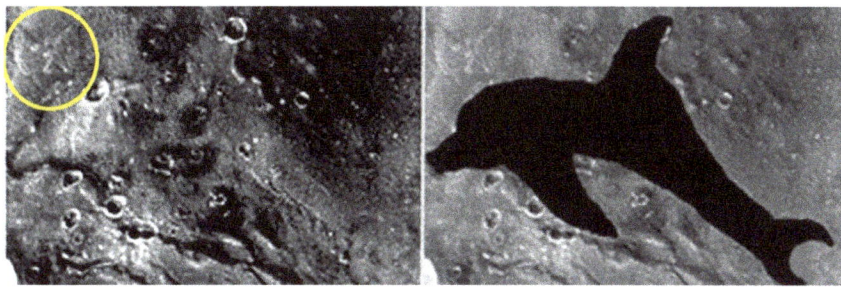

2.8 Left: Dolphin profile incised into the surface of Cydonia with the square and compasses markings immediately above its head. (MOC SP1-25803) Right: Highlighted by author.

Interestingly enough, it appears as though the architects on Cydonia were not the only ones to make a connection with the square and compasses symbol and the dolphin. This beautiful take of the square and compasses symbol, using a paddle for a compass and also incorporating a dolphin, is found in an insignia for at least 18 of the over 1800 Masonic lodges in Scotland (Figure 2.9).[9]

2.9 Masonic compasses and square insignia, Lodge Abercromby Provincial Grand Lodge of Stirlingshire Grand Lodge Of Scotland.

THE JAGUAR PROTECTOR

The Jaguar god is a key figure in the Mesoamerican religion and culture, manifesting in many entirely different personas. He is called Jaguar of the Underworld, Jaguar Sun (Lord Sun), Jaguar Night Sun, Baby Jaguar, and Water Lily Jaguar, and other names. Snarling jaguar heads appear atop the acropolis in Cerros and throughout the Maya and Olmec world. The jaguar was essentially an "uay," which is a divine manifestation of a human being or a god in animal form.[10]

A carved lintel panel from a temple in Tikal (Lintel 3 from Temple I at Tikal) depicts King Hasaw-Ka'an being carried on a palanquin. He wears a Sun god (Lord Sun) headdress and above him is the snarling "Jaguar god Protector" slashing his claws into the Otherworld[11] (Figure 2.10). On the arm of the Jaguar god Protector is a wristband with a split mask of a feline and a bird. The wing of the bird is composed of the fanned out "slashing claws" of the Jaguar Protector.

South of the Cydonia dolphin is a multifaceted structure using the artistic technique of contour rivalry. The top left side of this structure portrays a feline portrait, that of a left facing, snarling jaguar head we believe to be a depiction of the Jaguar Protector as seen on the Tikal lintel (Figure 2.10).

According to the Maya shamans, the "Jaguar Protectors," who were also called the "Balamob," were seen as sentinels. These jaguars were responsible for protecting the towns and fields of the kingdom in conjunction with the Sun god, Lord Sun.[12] The Balamob, or Jaguar Protectors, like the four sky-bears and rain gods, were all fourfold beings associated with the cosmos and the four cardinal points.

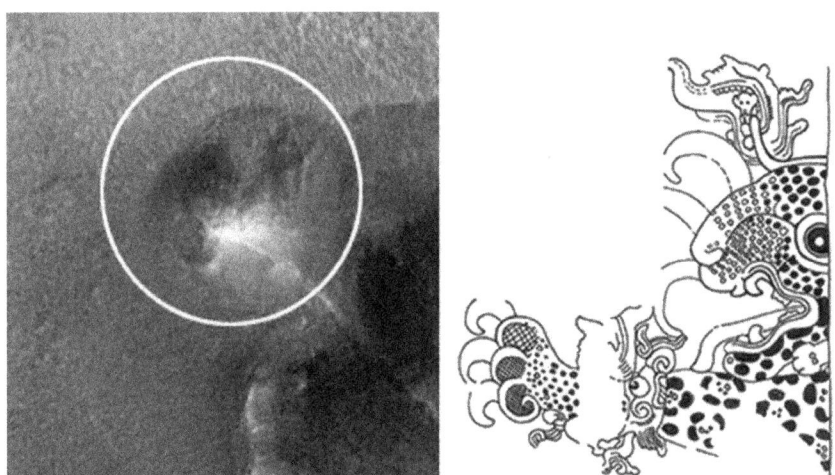

2.10 Left: Snaring Jaguar Mars; Right: Maya Jaguar Protector, lintel from Tikal. Drawing by George, J. Haas. (Image source: After John Montgomery).

THE COSMIC TURTLE

The turtle is seen as a symbol of the cosmos in many cultures. This primordial model is supported in the simple construction of its outer shell, which consists of a combination of round and square forms (compasses and square again). The Heavens are represented in the rounded top half of the turtle's shell (the carapace), the Earth in the square, flat bottom portion of the shell (the plasteron).[13]

The Cosmic Turtle in Maya mythology floated in the primordial sea of Creation, and it was from a small crack (or cleft) in its back that the First Father was reborn after being resurrected by his sons, the Hero Twins. As soon as First Father was reborn, a triangular hearth of three "throne stones" was set up by the gods on August 13, 3113 B.C.E. (The Maya calendar has been translated into our modern calendar).

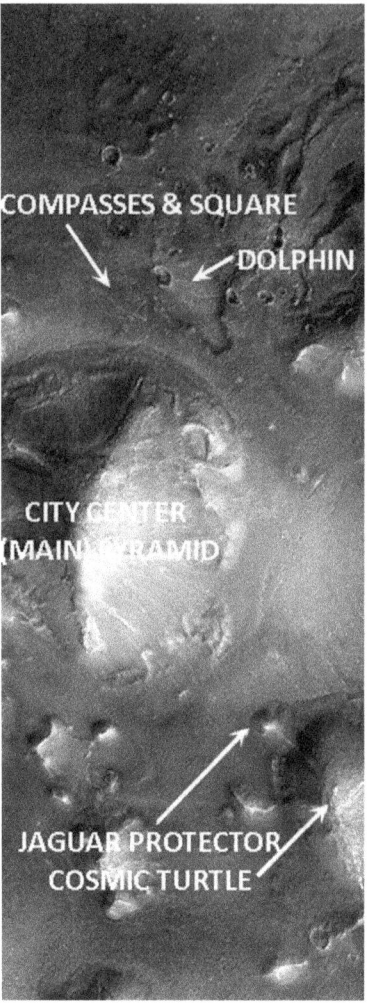

2.11 Location map.

August 13, 3113 B.C.E. is considered the starting date of the Maya calendar denoted in its western equivalent and the beginning of the 5th Sun or 5th age of Man. This triangle symbolized the "center of the new order" (the shape of the Greek letter "D," delta, is a triangle Δ and signifies a "door" or a "portal"). The hearthstones were also seen as the three stars of Orion's Belt, which were set across the back of the Cosmic Turtle.

The first of the stones was set in the form of the jaguar throne.[14] This jaguar throne acted much like the Jaguar Protector, guarding the cosmic center like a sentinel. Maya mythology states that five hundred and forty-two days after the resurrection of First Father through the cleft in the Cosmic Turtle, four gods set up the four sides and corners of the cosmos. They then erected the center tree, and together these signified the beginning of the "Fourth Creation," or "Fourth Sun."[15]

The image on the right in figure 2.12 is from a Maya stela that shows the Cosmic Turtle with a water-lily stem and bulb protruding from its shell. Since the water lily floated on the surface of the water, it was also a symbol for the division of the Middleworld and the Underworld from where the First Father was resurrected.[16]

On the same structure that displays the portrait of the Jaguar Protector (another aspect of the first of the three hearthstones), is just such a "Cosmic Turtle," complete with a water-lily bulb. Although the whole "turtle" structure was not captured by the Mars Orbital Camera in this third swath, the remaining data from the April 22, 2000 release of the right side (M09–05394) was added and a composite image was created (Figure 2.12).

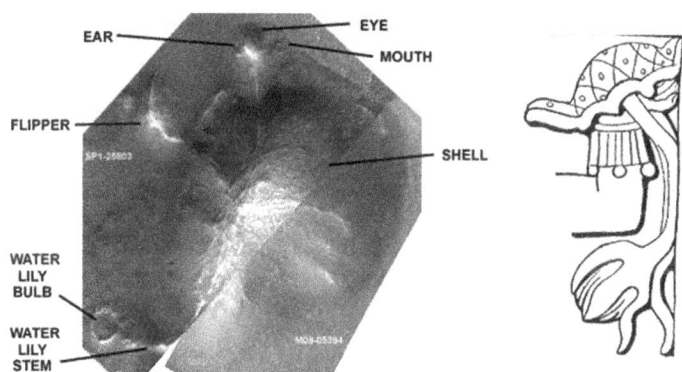

2.12 Left: Cosmic Turtle with water lily stem and bulb, Cydonia Mars (composite image of MOC SP1-25803 and MO9-05394). Right: Maya Cosmic Turtle with water lily stem and bulb. Drawing by George J. Haas. (Image source: Study of Maya Art, Spinden, page 18).

Once the image is completed, the "sea turtle" features become even more pronounced. A turtle "flipper" can be seen on the left side of the domed shell that exhibits the pronounced corrugated turtle-shell patterns around the top rim. A feature resembling the water-lily stem and bulb from the Madrid stela can be seen at the bottom left of the shell. The "bulb" of the water-lily plant is formed by the crater-like structure at the bottom of the stem.

If a mirror split is performed down through the center of the Mars Cosmic Turtle and continued along the "spine" of the turtle, the left half of the profile becomes a human-like face. The overall body of the structure forms a turtle shell with two "flippers" and a set of water-lily bulbs growing from the bottom of the turtle (Figure 2.13).

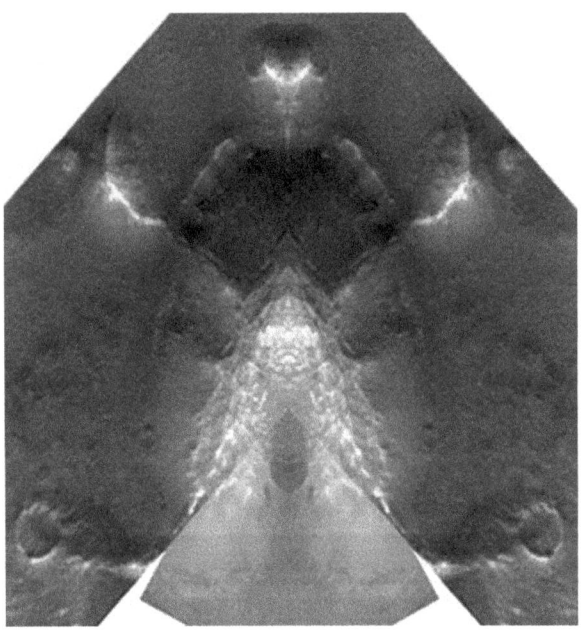

2.13 Left Side of Mars Cosmic Turtle Mirrored.

Located on one of the outer piers of the Lower Temple of the Jaguar at Chichen Itza is one of the most revealing images of Maya cosmology. Figure 2.14 depicts Pawahtun, one of the "old gods" who held up the sky at the time of Creation.[17] The facade features an image of Pawahtun standing with a turtle shell fastened to his chest. The turtle shell suggests that this "old god" is to be seen here as a personification of the Cosmic Turtle that floated on the surface of the primeval waters at the time of Creation. The Pawahtun god also wears a medallion with a "zero" sign, again signifying the "beginning time." A lily blossom can be seen hanging down in front of Pawahtun's headdress with a fish nibbling at its petals. A lily blossom and a fish are also featured on the Cydonia landscape.

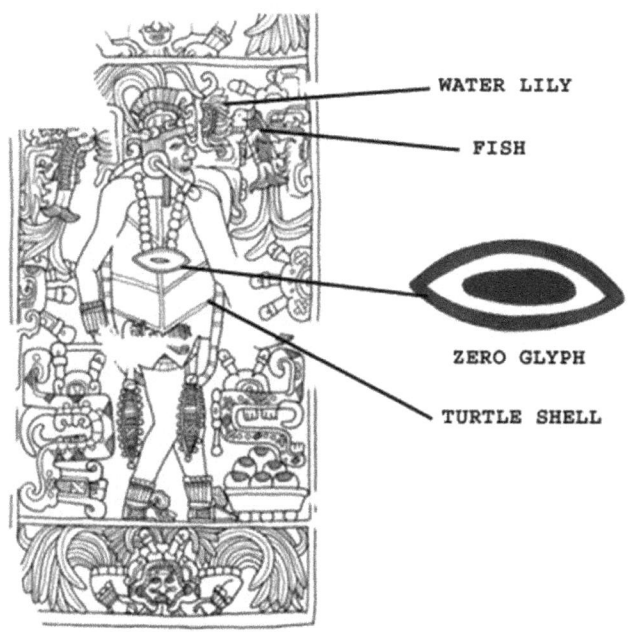

WATER LILY

FISH

ZERO GLYPH

TURTLE SHELL

2.14 Pawahtun Personifying The Cosmic Turtle on the Day of Creation; Outer Façade of The Temple of The Jaguar, Chichen Itza, Mexico; Note that a turtle shell forms the torso and he wears round ear flares (sun symbols and portals to the Otherworld) and a medallion with the Maya glyph for Zero.

A second depiction of Pawahtun, in the guise of a turtle, is seen in a Maya clay paint in figure 2.15. As you can see, when compared to the head of the mirrored Mars turtle, the two are almost identical.

When the right side of the Mars turtle is mirrored, a closed turtle shell is revealed with a turtle's head poking out of the top but appearing to be within the shell (Figure 2.16). This is in opposition to the left split, where the head and neck are noticeably protruding. Surprisingly, a floral pattern emerges from the center of the shell which forms an inverted, open "water lily blossom."

In Hinduism the symbol of a withdrawn turtle suggests "involution and return to the primeval state and therefore of a basic spiritual attitude."[18] This idea of the water lily and its connection to a spiritual birth and resurrection is also found within the Hindu religion in the form of the lotus, considered a close relative of the water lily.

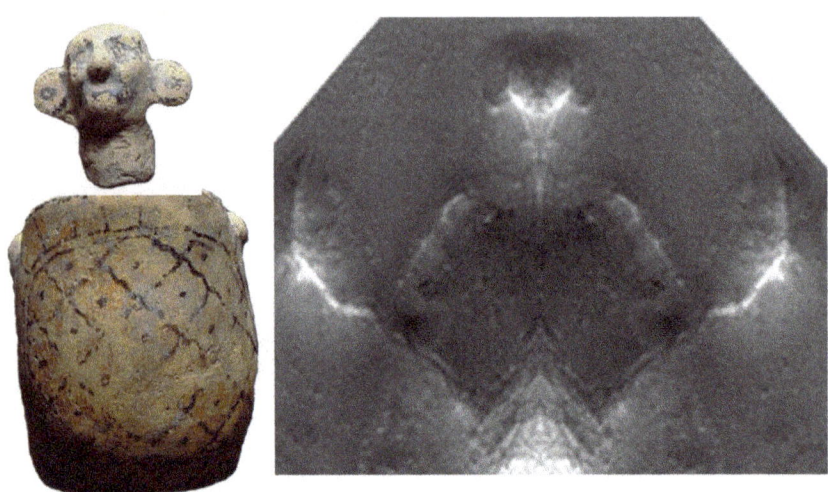

2.15 Left: Pawahtun with Turtle Shell, Maya Clay Paint;
Courtesy Justin Kerr. Right: Mirrored Mars Turtle Head.

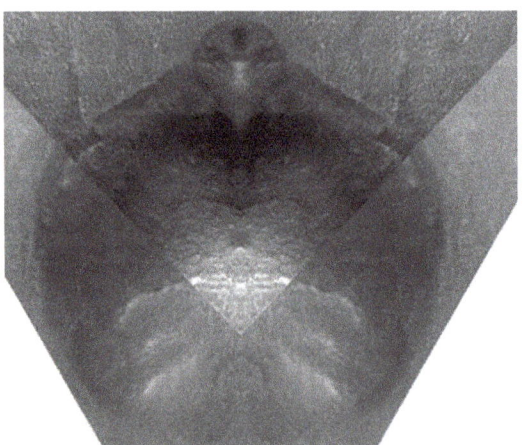

2.16 Mars Turtle With Indrawn Head.
Right Side Mirrored, note the open water
lily blossom on the lower shell.

THE "CITY CENTER PYRAMID" AND A TRINITY OF GODS OF THE FOURTH CREATION

Between the Dolphin and the Cosmic Turtle sits a structure which became known early on in the study of Cydonia as the Main Pyramid or City Center Pyramid (Figure 2.11). While spearheading a group of Mars researchers in the 1980's, Richard Hoagland proclaimed the structures in this area of Cydonia as remnants of a city. The structure is roughly 3 kilometers or 2 miles in size and has five ridges or spines which run from the top to a rounded, almost circular base. These ridges provide the structure with five sides or faces. This is not a symmetrical structure as one would expect of a pyramid. It does, however, provide us with five overlapping images and a sample of the complex and astounding art found in Cydonia.

The previous structure contained many facets of what the Maya called the Fourth Creation: The Jaguar Protector, The Cosmic Turtle, which floated on the Sea of Creation, and the Sky Bearer God, Pawahtun, that held up the sky at the time of Creation. Now we find, on this five-sided pyramid, the gods responsible for this Fourth Creation and its destruction.

The trinity of the Creation gods were seen as three separate gods, but they were ultimately viewed as one being. These three gods appeared in many forms with different visual features and names that occasionally overlapped, causing their specific aspects to become confusing.[19]

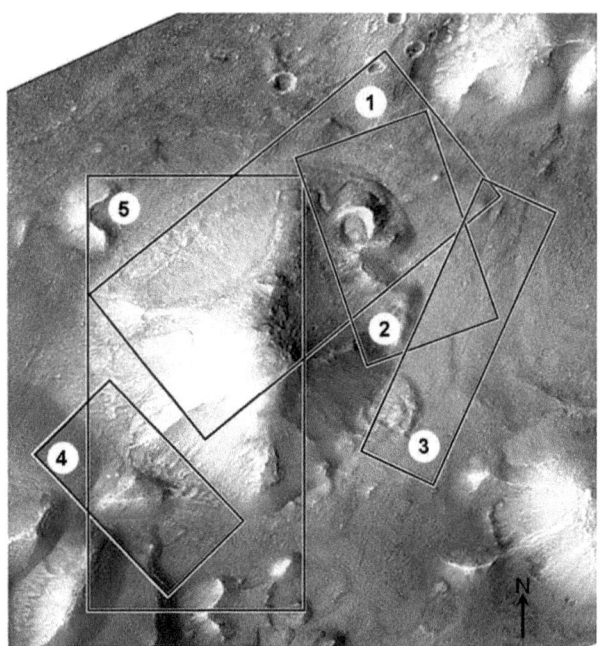

2.17 The Main or City Center Pyramid with context locations of 5 half pictographs. CTX: G22_026771_2213_XI_41N009W courtesy nasa/jpl/cal tech.

QUETZALCOATL (KET-SAHL-KOH-AHT-L)

Mesoamerica is the term used to describe Central America and the southern portion of North America. Archaeologists have determined there were two main civilizations that influenced the culture of this area. The Olmecs, believed to be the 'mother' culture, and the Maya. The two were undoubtedly contemporaries for a period of time. As one civilization falls, another develops to take its place, and such was the case in Mesoamerica with the Toltecs and Aztecs rising in later years. These civilizations also adopted or assimilated the Olmec and Maya cultural beliefs.

In one of the most important Maya temples, the Temple of Chac Mool, in Chichen Itza, Mexico, a sun disk was found carefully wrapped in a sacred bundle (Figure 2.18). The sun disk consists of a golden mosaic mirror of iron pyrite surrounded by a brilliant turquoise mosaic version of the sun disk divided into eight compartments.[20] The profile of a feathered serpent occupies every second compartment. If the profile image of the one of the serpents is mirrored, it produces a frontal presentation of a feathered serpent (Figure 2.19).

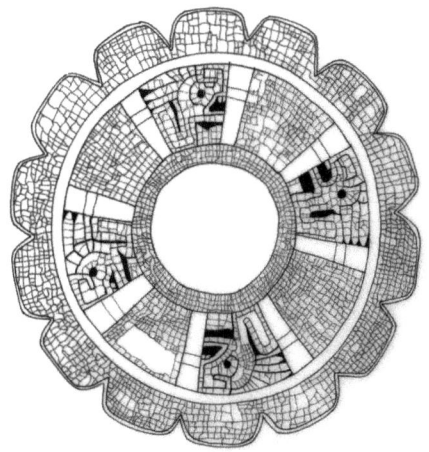

2.18 Mosaic Sun Disk from the Temple of Chac Mool showing a circle of feathered serpents with a mirror in the center; Drawing by Linda Schele, copyright David Schele (Courtesy Ancient Americans at LACMA).

2.19 Profile Feathered Serpent when mirrored.

The most important god in the pantheon of the Maya, the Feathered Serpent, was known by many names. Gucumatz to the Quiche Maya, Kukulkan to the Yucatan Maya, and most famously, in the Nahuatl language, as Quetzalcoatl. Worshipped as early as the first century B.C.E. at the great city of Teotihuacan, Quetzalcoatl was a combination of two words. Quetzal is a sacred bird of the Yucatan and coatl means serpent or twin. This god is identified as one of thirteen deities who shaped the world and created human beings. He was also said to be the Creator of the Fifth Sun or the Fifth Age of Man. From him, humans learned the rules of law, agriculture, literacy, the arts, medicine, architecture, construction, hunting, fishing, and all other aspects of civilization.[21] The epithet referred to the duality of a bird and a serpent. He was the Creator god and the son of First Mother and First Father. There were many other aspects to this deity as he was also considered the embodiment of Venus (First Lord) and the Sun (Jaguar Sun), the two representations found in the split-faced 'Face on Mars' in chapter 1.

The pictograph labeled 1 in figure 2.17 is roughly 2 kilometers in size. Haas and I believe this to be a presentation of a half image of Quetzalcoatl. Our determination that this is a portrayal of Quetzalcoatl is not just due to the accompanying bird and twin serpents, but the similarity to a depiction of him from an Aztec image contained in the Codex Telleranio (Figure 2.20). In this image, a helmeted Quetzalcoatl wears serpent braids and a plumed headdress.

2.20 Quetzalcoatl; From Codex Telleranio.
Drawing by George J. Haas.

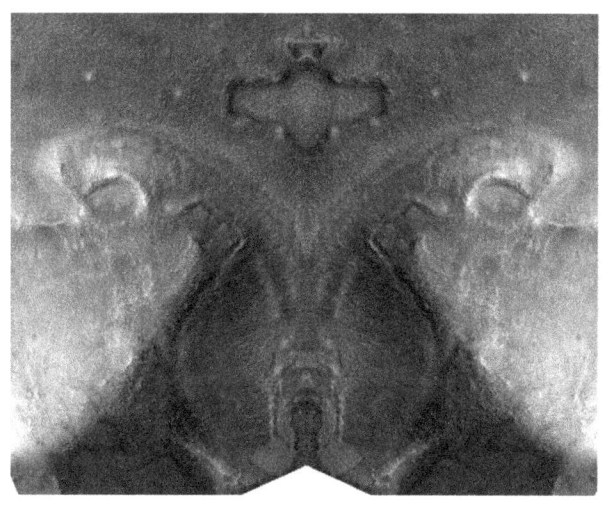
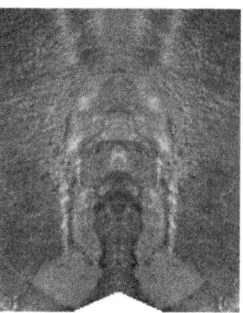

2.21 Quetzalcoatl mirrored image. Note the twin serpents rising out of the helmet, the striations and the depiction of billows of smoke emitted from the mouth of the serpents. Note also the bird/flying craft above, which is formed from the head of the incised, bottle-nosed dolphin.

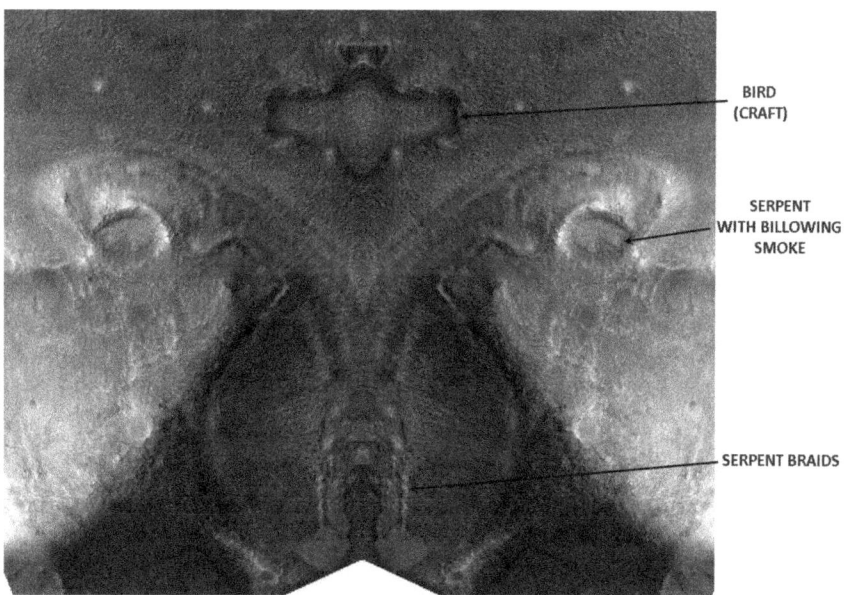

BIRD
(CRAFT)

SERPENT
WITH BILLOWING
SMOKE

SERPENT BRAIDS

2.22 Quetzalcoatl, The Feathered Serpent, mirrored, on Mars. False color added by author.

LORD SUN (THE ADMIRAL)

The half pictograph marked #2 in figure 2.17 is roughly 2 kilometers in size, and when mirrored, forms a striking portrait of a humanoid with a face that presents a somewhat feline affectation. He wears a uniform adorned with a high broached collar and a large shield-like breastplate (Figure 2.23). This breast plate is formed from the same feature as the serpent head in the Quetzalcoatl image. This strange character has small squinting eyes, what may be a centered T-shaped tooth, and a nose defined by an intricately designed nose ornament consisting of the letter M. The highly reflective nature of the surface surrounding his head gives an aura-like halo, most likely representing the Sun.

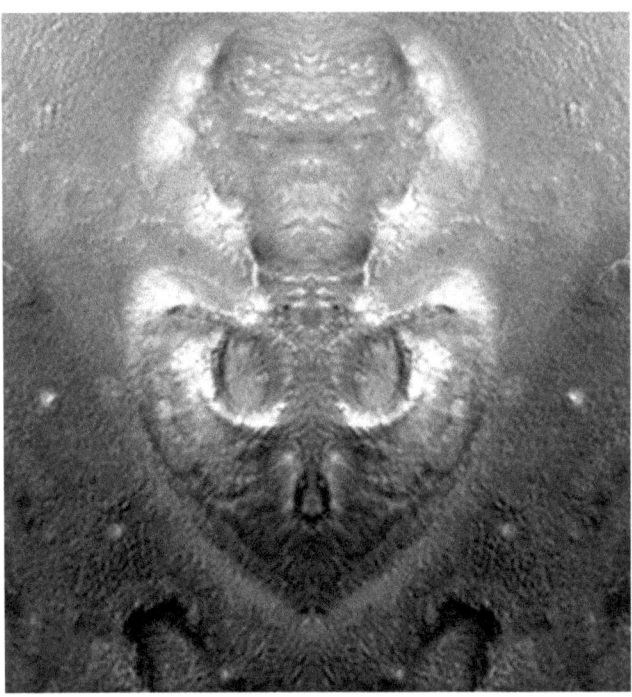

2.23 Mirrored image of Lord Sun (the "Admiral"). MOC SP1-25803 Courtesy NASA/JPL/CAL TECH.

2.24 Lord Sun (Maya Jade Head).
Note the bound flank of hair, the tooth, the
m-shaped nose and the long ears.

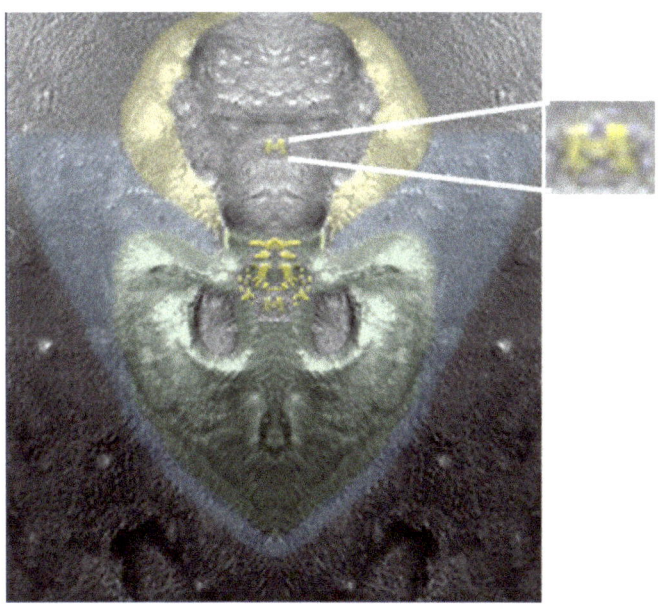

2.25 Mirrored half image, notice the letter M nose ornament and
the halo. Also notice the lighter colored surface (sky) at the top, and
the darker surface (sky) below, indicating the rising of the Sun.
False color added by author.

Halos were part of early art work defining the subject as "holy" or a "deity". They were symbols of light but can traced back in history to represent the Sun. They can be found in the art work of ancient Egypt and the religious artwork of Buddhism, Hinduism and Judeo-Christianity.

2.26 Religious figures with halos.

Before we recognized this image as the Maya Lord Sun, I was so taken by the formal demeanor of this genteel character that I called him "The Admiral," a nickname that would prove very prophetic.

After an extensive examination of this image, Haas and I eventually decided that it actually represents an aspect of the Maya Jaguar god also known as Lord Sun (Figure 2.24). The Maya Sun god, (Ahau-kin Lord Sun) is identified most commonly as a human-jaguar hybrid with a gathered lock of hair on his head, a T-shaped tooth, and squinting eyes.[22] Mythologist Douglas Gillette gives this poetic description of the Jaguar god in his book *The Shaman's Secret: The Lost Resurrection Teachings of the Ancient Maya:*

> *"When the Underworld slid up above the Earth at night to become the star-studded sky, the patterns of the constellations were thought to be the Jaguar's hide, now seen in all its diabolic immensity. But the Jaguar god also bore in his terrible darkness the promise of resurrection and rebirth for the living and the dead, for he was also the Underworld Sun— more golden in the Underworld than black, and destined to rise at dawn as the Lord of Life."[23]*

The Maya viewed Lord Sun as only one aspect of the many incarnations of the Jaguar god. Lord Sun was also part of the "triad" of Creation at Palenque.

We believe the image in Figure 2.23, also represents the Sumerian god Enki, who is the god of Water and of Creation. Interestingly, noted author and historian Zecharia Sitchin believes Enki was also the Egyptian Creator god Ptah. Furthermore, we believe that Enki was ultimately the prototype of the Maya trio of First Father, First Lord and First Jaguar.

The basic iconography shared by all these gods leads us to suggest that they may have been one and the same. There are two major clues that provide us with this conclusion and may explain the complex threefold aspects of Lord Sun as a humanoid, jaguarian, and amphibious Sun god. The first piece of evidence is presented in Sitchin's book, The Cosmic Code.

> *The secret numbers of the gods can serve as clues to the deciphering of secret meanings in other divine names. Thus, when the alphabet was conceived, the letter "M"—Mem, from Ma'yim, water, paralleled the Egyptian and Akkadian pictorial depictions of water (a pictograph of waves) as well as the pronunciation of the term in those languages for "water." Was it then just a coincidence that the numerical value of "M" in the Hebrew alphabet was "40"—the secret numerical rank of Ea/Enki, "whose home is water," the prototype Aquarius?* [24]

According to Sitchin, this ancient numbering code reveals the numerical placement of the Sumerian gods. The gods are ordered as follows: Anu, the father of the gods, is numbered 60; Enlil, who is Lord of Earth, is numbered 50; Ea/Enki, the Water god, is numbered 40; Sin, the Moon god, is numbered 30; and Shamash, the Sun god, is 20. [25]

In examining the portrait of the Martian version of Lord Sun, it is hard not to miss the nose ornament, a very prominent letter "M" (Figure 2.22, 2.24). The Maya had a similar encoding system to the Sumerians. In the Mayan language the numbers are expressed, not only by the traditional "bar and dot" method, but also by individual portraits of their gods. These numerical profiles, of which there are 20 different heads, are called "head variants" (Figure 2.27).

The Mayan head variant for the number 13 is seen as a profiled head of a water deity, who is a personification of large bodies of water, such as lakes and oceans. In our modern alphabet the letter "M" is the thirteenth letter and, as explained by Sitchin, its design is based on an ancient pictograph depicting waves, another water sign.

The Mayan head variant for the number 4, which represents Lord Sun, can be seen in figure 2.27. It is very interesting that this idea of numbered gods was shared by such diverse cultures as the Sumerians and the Maya.

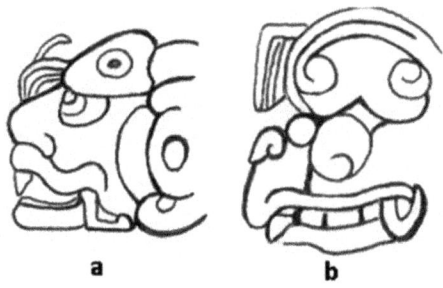

2.27 Maya Number Glyphs
a.) Lord Sun - Head Variant for number 4.
b.) Water Deity - Head Variant for number 13.
Drawings by George J. Haas (Image source: *Lost Worlds; the Romance of Archaeology*, White, 249).

Also, it appears they are sharing the image of their god on Mars. The Sumerian Creator god Enki was commonly depicted surrounded by sea creatures and, as we can see, this Martian image is as well (Figure 2.28). So the nicknaming this image as the Admiral early on, was indeed rather prophetic.

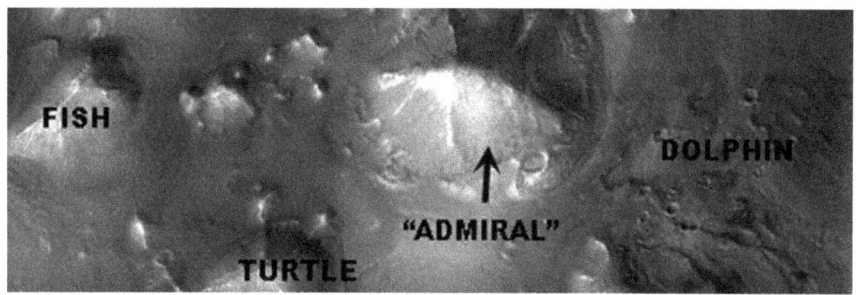

2.28 Location of Lord Sun, the "Admiral", surrounded by sea creatures. (MOC SP1-25803)

THE WATERLILY JAGUAR

The Water Lily Jaguar is centered immediately above Lord Sun (the Admiral) in figure 2.17 and is number 3 in the series of 5 images. This half image when mirrored, becomes that of a feline believed to be a tiger in a sitting position (Figure 2.28). This figure is roughly two and a half kilometers from head to toe (one and one half miles). The first oddity we noted in this pictograph was that the squatting feline at Cydonia is depicted with a strange "tie" or "scarf" hanging from its neck. It also has a headdress and a feature resembling a headband that sweeps across its forehead. Crowns and headdresses were depicted on both humans and animals throughout Mesoamerican art and were cultural markers of divinity. An odd floral-like feature, resembling a water lily ornament, sits upon the feline's head. This particular ornament reminded us of the Maya Water Lily Jaguar. In Maya iconography the normal attire of the Water Lily Jaguar (another form of the Jaguar god), is a knotted scarf worn around his neck and a water-lily sign featured on his brow or atop his head. The scarf is a symbol of sacrifice, and the water-lily sign, a metaphor for royal power, can take the form of a leaf, a flower, or a net-like pattern as seen on the shell of turtles.

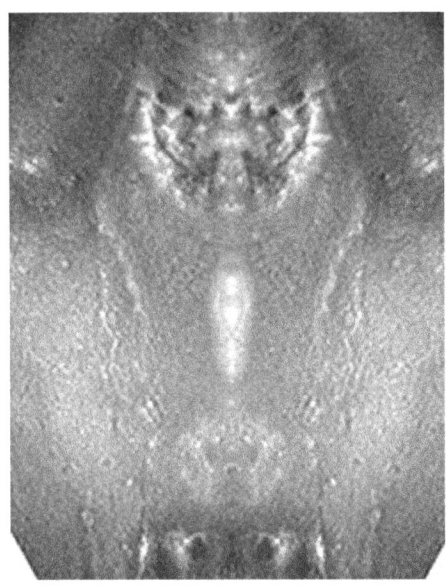

2.29 Mars Tiger
Mirrored image of Water Lily Jaguar, (MOC SP1-25803).

An ancient urn, found in Monte Alban, Mexico, is sculpted in the shape of a squatting, snarling jaguar, very much like the Cydonia tiger. The vessel dates back to about 200 B.C.E. and is 91 centimeters (36 inches) in height (Figure2.30).[26] Notice the knotted scarf that hangs around its neck which, as stated, is a symbol of sacrifice. Perhaps that is why the Zapotec Jaguar urn was found in a cemetery in Monte Alban, Mexico. Upon closer examination, the "scarf" feature of the Cydonia tiger has, within its amphora design, what appears to be the image of an open-tailed fish, much like the common Pisces symbol of the zodiac and the one used by Christians to symbolize the resurrected Christ. In the eighteenth Sura of the Muslim holy book, the Koran, the fish is seen as a symbol of resurrection.

One has to wonder why the Water Lily Jaguar in Cydonia has the face of a tiger. As stated, the Water Lily Jaguar was a denizen of the Underworld where souls went before being resurrected, but was also an aspect of the Jaguar god who was also known as Balam-U-Xib or Jaguar Moon Lord.[27] The tiger was also an ancient Chinese symbol of the New Moon, representing resurrection.

2.30 Mesoamerican crouching Jaguar shaped Urn. Note the scarf. (Image source: Photograph by Xuan Che, Oaxaca Urn Jaguar, Symbol of Monte Alban, the Zapotec capital. Museo Nacional de Antropología, Mexico City. Creative Commons – Attribution 2.0 Generic – CC BY 2.0)

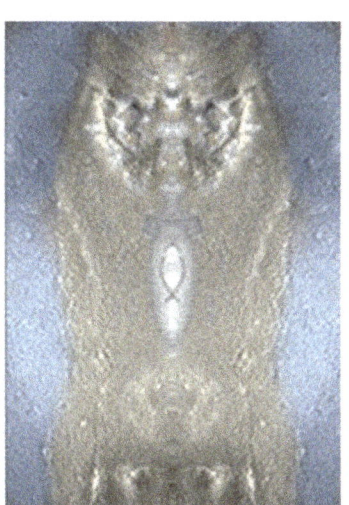 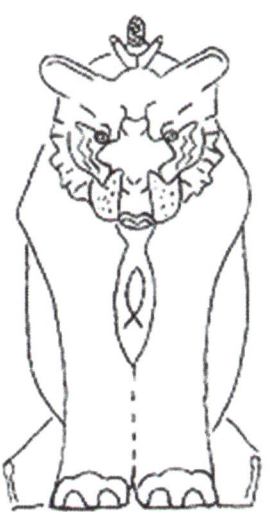

2.31 Left: "Water Lily Jaguar on Mars, false color. Right: Drawing by George J. Haas. Note the floral symbol on the head and the fish symbol on the scarf.

THE VISION SERPENT AND FIRST FATHER

The Vision Serpent was another aspect of Quetzalcoatl. As you remember, the Feathered Serpent was known as Gucumatz to the Quiche Maya, Kukulkan to the Yucatan Maya, and most famously, in the Nahuatl language and to the Aztecs, as Quetzalcoatl. However, he was also known as the Vision Serpent to the southern Maya.[28]

The Vision Serpent acted as a portal which connected the mortals to the immortals (gods) and a gateway to the spirit realm and ancestors. When the Vision Serpent was summoned a spirit or deity would appear in the open mouth of the serpent (Figure 2.32).

The fourth half image on the Main Pyramid from figure 2.17, when mirrored, depicts a person emerging from the mouth of a Vision Serpent. Like the nose ornament on the Lord Sun image, we see another letter 'M' above the head of the serpent, and twin serpents, a symbol of Quetzalcoatl, below the shoulders of the person. We believe this spirit or ancestor represents First Father. As we stated, First Father, First Lord and First Jaguar are believed to be aspects of the same being. So here, on this large structure just above the Cosmic Turtle which floated on the Sea of Creation we have, the three aspects of the Maya Creation Triad.

2.32 Maya Vision Serpent with spirit or ancestor appearing out of the serpent's open mouth. Drawing by George J. Haas (image source: A Forest of Kings, by Schele and Freidel, figure 2:3, page 69).

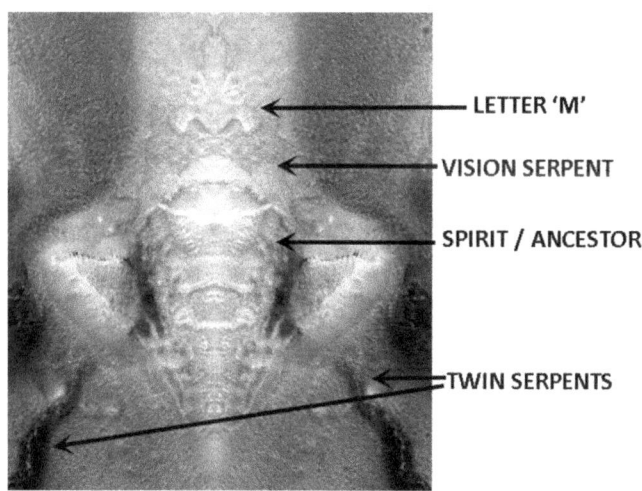

LETTER 'M'

VISION SERPENT

SPIRIT / ANCESTOR

TWIN SERPENTS

2.33 Mirrored image of Vision Serpent With Spirit / Ancestor Emerging from Mouth.

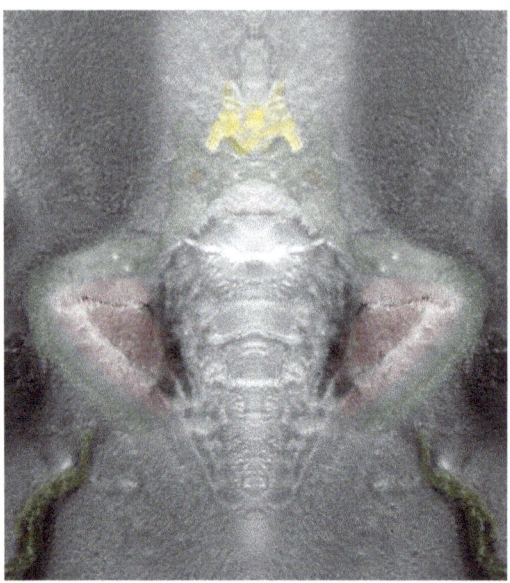

2.34 Mirrored image of Vision Serpent with First Father appearing in the open mouth. Notice the twin serpents on the arms and shoulders and the 'M' above the head! Color added by author.

THE RIVER OF RESURRECTION

This entire scene is quite astounding and took much time to solve. Half image number 5 from figure 2.17 is nearly 4.5 kilometers or over 2.5 miles from top to bottom and is oriented north – south.

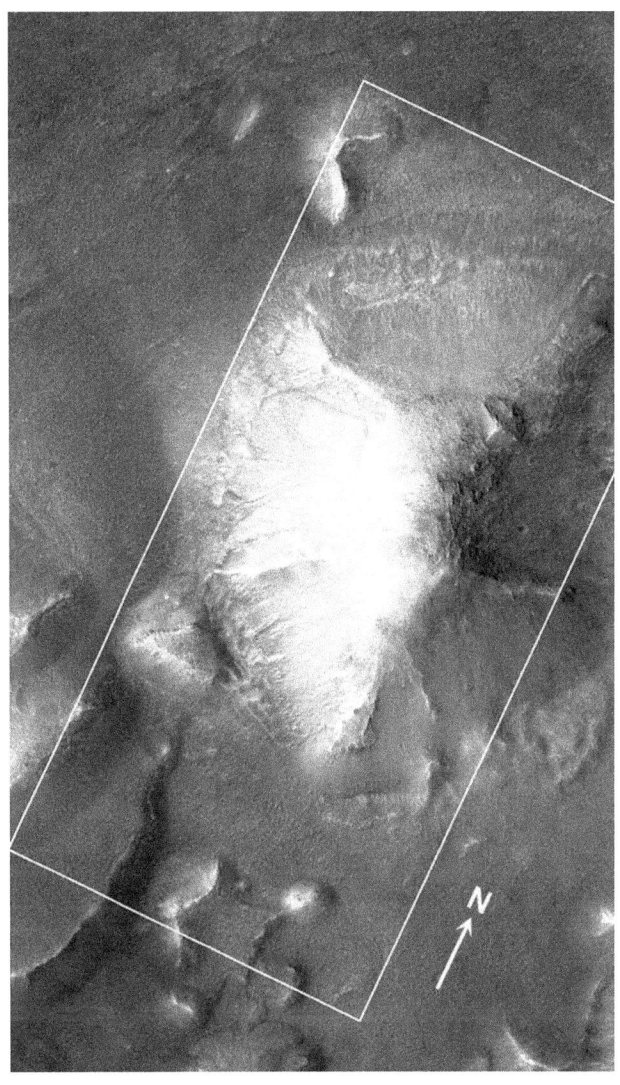

2.35 The River of Resurrection and Creation half image location. Cropped from MOC E02-01847.

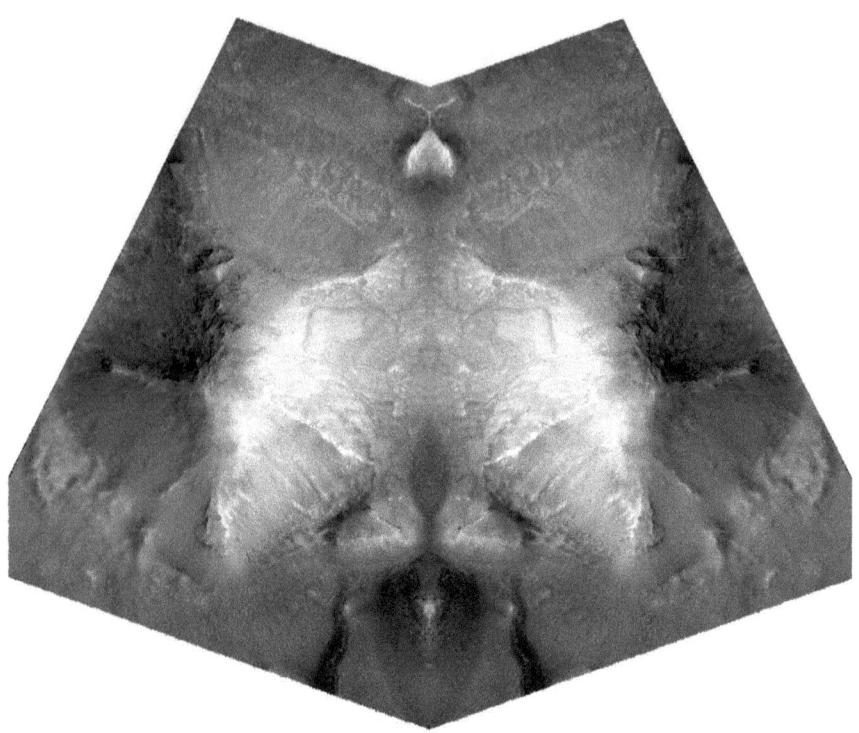

2.36 The River of Resurrection. Mirrored image of figure 2.35.

When mirrored, the half image becomes a complex scene. A woman, wearing a shear nursing blouse, has her hands placed upon a tree stump as she looks down. Water flows from her breasts and runs down into a small stream and waterfall. Her breasts rest upon the head of a bearded man in an almost comical fashion, pressing his ears down. He crouches in the stream, tongue out, apparently lapping the water. Just below his tongue is a frog. Immediately below the frog is a fish dropping down the waterfall. A second fish appears at the bottom of the waterfall, holding a piece of jewelry in its mouth, consisting of a five-pointed star. Lastly, beside the fish, are twin serpents.

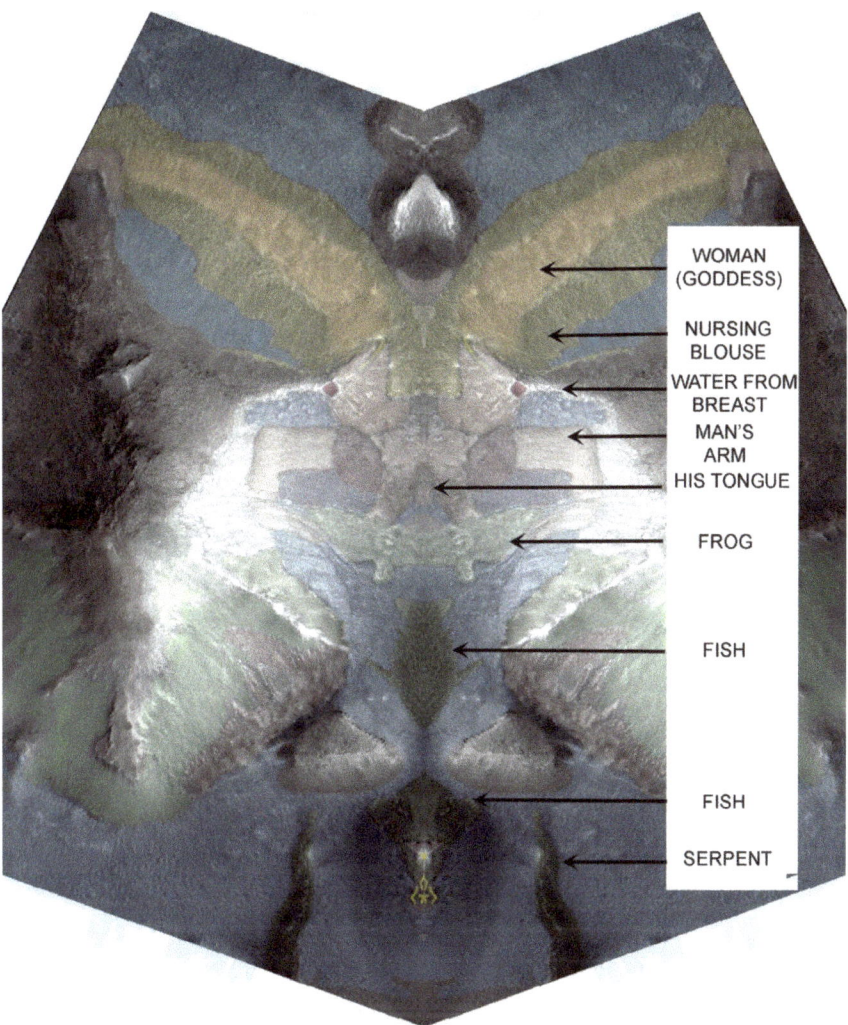

2.37 Mirrored Image of The River of Resurrection and Creation. Color added by author.

Let us see how this scene translates in Mesoamerican mythology. Although this structure contains the triad of the 4th Creation, in Aztec mythology, it was the goddess Chalchiuhtlicue that both oversaw the Creation and caused the destruction of the 4th Sun.

Chalchiuhtlicue is the Aztec Goddess of water, oceans, lakes, rivers and springs. Her name means "Woman of the Jade Skirt," and she is often depicted with water-lilies, dressed in watery blues and greens, and sometimes has quetzal-feathers in her hair. In Aztec mythology, this world has seen five Suns or Creations. Chalchiuhtlicue presided over the Fourth Sun, or the Fourth Creation of the world, and its destruction.[29] Similar to the Hebrew story of God flooding the world with forty days and forty nights of rain, she brought about its destruction by releasing fifty-two years of torrential rains. However, she also protected humanity by changing the people into fish so that the waters would not drown them and by creating a bridge linking Earth to Heaven for the souls of those she favored. In the mid-19th century, archaeologists unearthed a 20-ton monolithic sculpture depicting a water goddess that is believed to be Chalchiuhtlicue from underneath The Pyramid of the Moon.[30]

It is believed that she is an equivalent goddess to the Maya Ix Chel, more recently known as Chac Chel. In this depiction of the Maya Moon goddess from the Dresden Codex she is shown with water cascading from her breasts. Like her Aztec counterpart she is identified with world ending floods.[31]

2.38 Maya Moon goddess, from Dresden Codex.
Shows water flowing from her breasts and blood
from skirt. Drawing by George J. Haas

Just below the goddess' breasts is the depiction of a man submerged to his chest and lapping the water. In Aztec tradition, a midwife would say a prayer to Chalchiutlicue: "Behold this element without whose assistance no mortal being can survive." She also would sprinkle water on the breast of the baby while saying, "receive this celestial water that washes impurity from your heart." Then she would go to the head and say, "son receive this divine water, **which must be drank** that all may live, that it may wash you and wash away all your misfortunes, part of the life since the beginning of the world: this water in truth has a unique power to oppose misfortune." Finally, the midwife would wash the entire body of the baby and say, "in which part of you is unhappiness hidden? Or in which part are you hiding? Leave this child, today, he is born again in the healthful waters in which he has been bathed, as mandated by the will of the (goddess) of the sea, Chalchiutlicue."[32]

Below the man lapping the water is a caricature of a frog. The frog, according to J.E. Cirlot and the *Dictionary of Symbols*, is also a lunar animal as well as an attribute of the Egyptian goddess Heqet, the goddess of fertility and childbirth. [33] According to 19th century philosopher and author Helena Blavatsky, **the frog was one of the principle creatures associated with the idea of creation and resurrection**.[34] Cirlot also states that **the frog represents the transition from the element of the earth to that of water** and vice versa.

Five of the twenty big celebrations in the Aztec calendar were dedicated to Chalchiutlicue and her husband (or brother), Tlaloc. **During these celebrations, priests dove into a lake and imitated the movements and the croaking of frogs,** hoping to bring rain.

Below the frog is a fish descending a waterfall with a second fish in the pool at the bottom holding what looks like a piece of jewelry consisting of a five-pointed star. As was stated, before **Chalchiutlicue destroyed civilization with a great flood, she turned the people into fish** so they would not drown. The fish itself is a symbol of resurrection, not only to the Maya, but also in Christian and Islamic mythology. The five pointed star also has numerous aspects to its symbolism. Once again Cirlot states:

> *"the star is a symbol of the spirit, however it rarely carries*
> *a single meaning. It nearly always alludes to multiplicity*

in which case it stands for the forces of the spirit struggling against the forces of darkness and this meaning has been incorporated into emblematic art all over the world. The five pointed star in particular, as far back as the ancient Egyptian hieroglyphics, signified 'rising upwards towards the point of origin'"…[35]

According to the legend, Chalchiuhtlicue's destruction by flood made way for the 5th Creation or 5th Sun. Does this five pointed star in the mouth of the fish also signify the Fifth Sun of Creation?

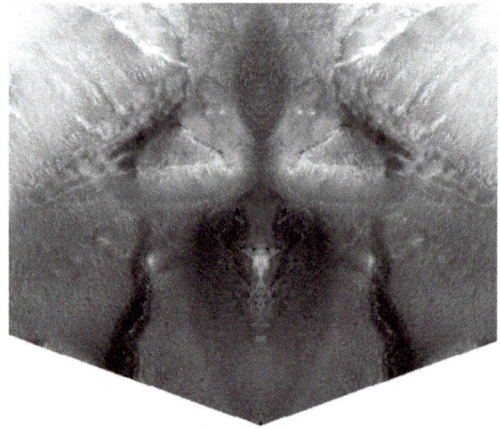

2.39 Detail of figure 2.37 showing the two fish with five pointed star.

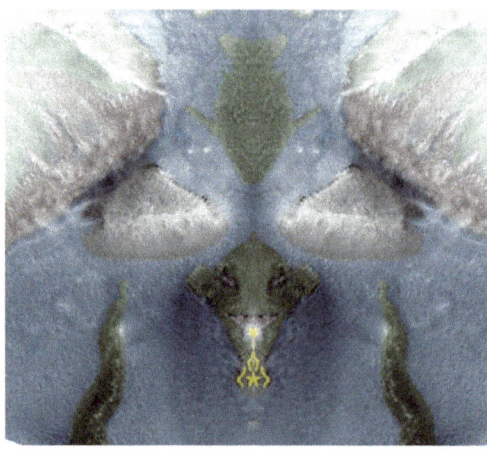

2.40 Fish with five pointed star. (Color added by author).

FOOTNOTES:

1. Delia Goetz and Sylvanus G. Morley (English version) from the translation of Adrian Recinos, *Popol Vuh; The Sacred Book of the Ancient Quiche Maya* (Norman, University of Oklahoma Press, 1950), 80.
2. Frank W. Porter III, *The Maya* (New York: Chelsea House, 1991), 59.
3. Adrian G. Gilbert and Maurice M. Cotterell, *The Mayan Prophecies: unlocking the secrets of a lost civilization* (Rockport: Element, 1995), 68-84.
4. Linda Schele and Peter Mathews, *The Code of Kings: The Language of Seven Sacred Maya Temples and Tombs* (New York: Touchstone, 1999), 111, 112.
5. Kline, Ecuador, 330
6. J.E. Cirlot, *A Dictionary of Symbols* (New York: Barns & Noble, 1995), 46, 307.
7. Ibid. 61.
8. Peter Tompkins, *Mysteries of the Mexican Pyramids; Dimensional analysis on original drawing by Hugh Herleston, Jr. And Historic Illustrations from many Sources* (New York: Perennial Library, 1976), 282, 283.
9. http://www.standrew518.co.uk/Address/View_all.php#bookmark
10. Douglas Gillette, *The Shaman's Secret; The Lost Resurrection Teachings of the Ancient Maya* (New York: Bantam, 1997), 229. The word "uay" is also spelled "way" and is pronounced "why."
11. Linda Schele and David Freidel, *A Forest of Kings, The Untold Story of The Ancient Maya* (New York, Quill 1990), 210.
12. David Freidel, Linda Schele, and Joy Parker, *Maya Cosmos; Three Thousand Years on the Shaman's Path* (New York: Quill, 1993), 50.
13. Jean Chevlier and Alain Gheerbrant, *Dictionary of Symbols* (New York: Penguin, 1996), 1016.
14. Linda Schele and Peter Mathews, *The Code of Kings: The Language of Seven Sacred Maya Temples and Tombs* (New York: Touchstone, 1999), 37, 217.
15. Ibid., 37.

16. Linda Schele and David Freidel, *A Forest of Kings, The Untold Story of The Ancient Maya* (New York, Quill 1990), 209.

17. Linda Schele and Peter Mathews, *The Code of Kings: The Language of Seven Sacred Maya Temples and Tombs* (New York: Touchstone, 1999), 214.

18. Jean Chevlier and Alain Gheerbrant, *Dictionary of Symbols* (New York: Penguin, 1996), 1019.

19. Linda Schele and Mary Ellen Miller, *The Blood of Kings: Dynasty and Ritual in Maya Art* (New York: George Braziller, Inc., 1985), 48.

20. Linda Schele and David Freidel, *A Forest of Kings, The Untold Story of The Ancient Maya* (New York, Quill 1990), 394.

21. Ancient History Encyclopedia, https://www.ancient.eu/article/415/the-mayan-pantheon-the-many-gods-of-the-maya/

22. Mary Ellen Miller, *Maya Art and Architecture* (New York: Thames & Hudson, 1999). 74.

23. Douglas Gillette, *The Shaman's Secret; The Lost Resurrectrion Teachings of the Ancient Maya* (New York: Bantam Books 1997), 92.

24. Zecharia Sitchin, *The Cosmic Code*: Book VI of The Earth Chronicles (New York: Avon, 1998), 172.

25. Ibid: 171.

26. Jonathan Norton Leonard and the Editors of Time-Life Books, Great Ages of Man; A History of the World's Cultures: Ancient America (New York: Time Inc., 1967), 30.

27. Douglas Gillette, *The Shaman's Secret; The Lost Resurrectrion Teachings of the Ancient Maya* (New York: Bantam Books 1997), 92.

28. Linda Schele and David Freidel, *A Forest of Kings, The Untold Story of The Ancient Maya* (New York, Quill 1990), 394.

29. Karl Taube, *Aztec and Maya Myths* (University of Texas Press, 1993), 34, 35.

30. https://en.wikipedia.org/wiki/Chalchiuhtlicue#cite_ref-11(Berlo 1992: 138; Pasztory 1997), 87–89

31. Vail and Hernandez, *Rain and Fertility Rituals in Postclassic Yucatan Featuring Chaakand and Chak Chel". The Ancient Maya of Mexico: Reinterpreting the Past of the Northern Maya Lowlands*, 2014.

32. de Sahagún, Bernardino Florentine Codex: *General History of the Things of New Spain, Book 6: Rhetoric and Moral Philosophy.* (School of American Research, 1970). 175

33. J.E. Cirlot, *A Dictionary of Symbols* (New York: Barns & Noble, 1995) 114.

34. Ibid. 114.

35. Ibid: 309, 310.

CHAPTER III

CYDONIA, MARS AND THE STAR MAP OF GREATER ORION

"We are all ... children of this universe. Not just Earth, or Mars, or this system, but the whole grand fireworks. And if we are interested in Mars at all, it is only because we wonder over our past and worry terribly about our possible future."

— Ray Bradbury,
Mars and the Mind of Man, 1973.

THE CROWNED LION

On June 2, 2003 the European Space Agency launched the Mars Express Spacecraft. The spacecraft had a relatively short flight, as it took advantage of the orbits of Earth and Mars, which brought the planets as close as they had been in about 60,000 years. The spacecraft was inserted into Mars orbit on December 20, 2003. With the European Space Agency becoming involved in the exploration of the planet Mars, we got our first crisp, regional images of Cydonia. This regional view is roughly 72 kilometers by 87 kilometers and has west at the top and north to the right.

What many viewers noted upon first viewing this image was the appearance of the gigantic face of a lion. Upon closer examination, I found the lion was wearing a three-point crown, the Maya symbol of royalty, as we had found with the mirrored humanoid side of the Face on Mars. The lion's gaze seems to be focused on a crater in front of its face.

Across from the lion lie three more craters. These craters demonstrate a strong correlation to the positioning of the pyramids on the Giza Plateau in Egypt, and by extension, to the positioning of the three stars of Orion's belt (Figure 3.3).

If one overlays a star map, the positions of the three Orion belt stars match the craters. The crater size, however, does not match the intensity scale, indicated by the diameter of the circle. Could this signal that they are not in a normal position relative to the other stars of the Orion constellation when viewed in a 2-dimensional presentation? The alignment of the three craters points to another, larger one. (Figure 3.4)

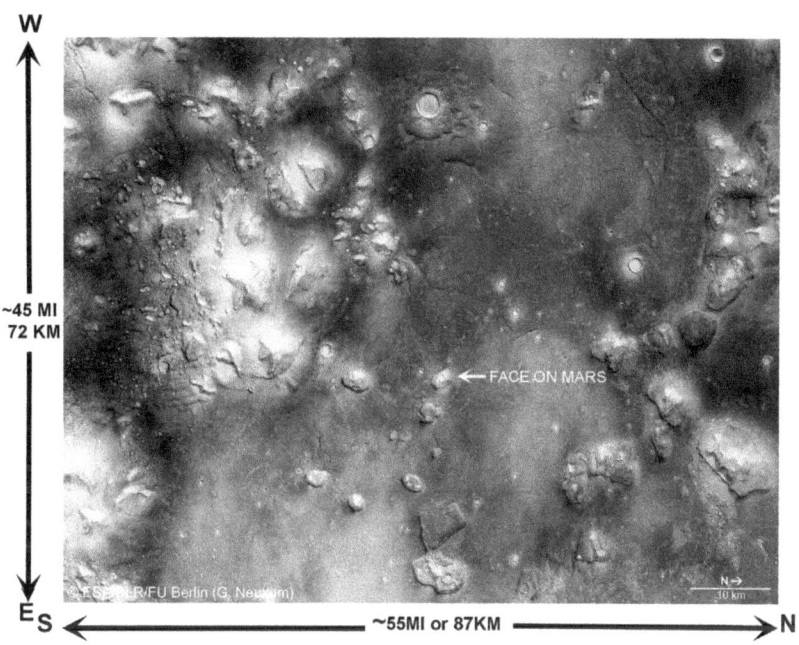

3.1 300-230906-3253-6-nd1 Cydonia, courtesy of European Space Agency.

3.2 Lion head, wearing three point crown, focused on crater marked in red. False color added by author. 00-230906-3253-6-nd1 Cydonia, courtesy of European Space Agency.

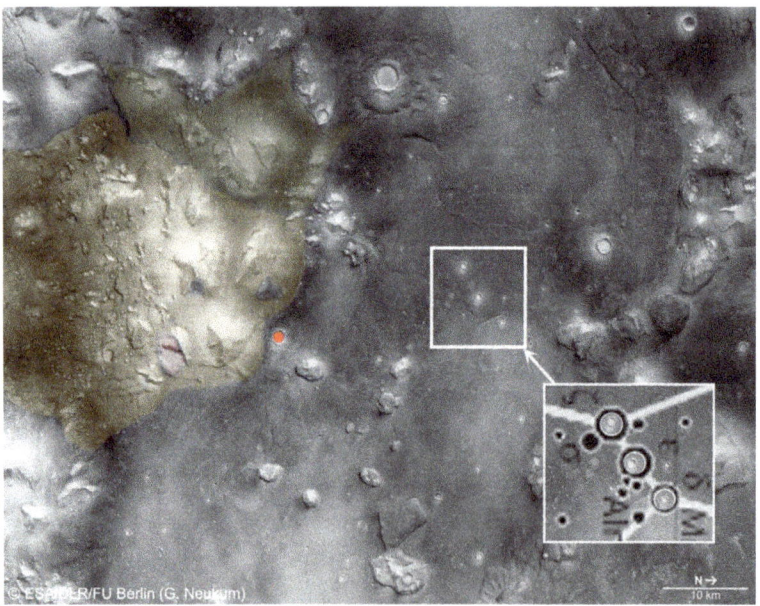

3.3 Highlighting of the 3 craters overlain with Orion belt stars.

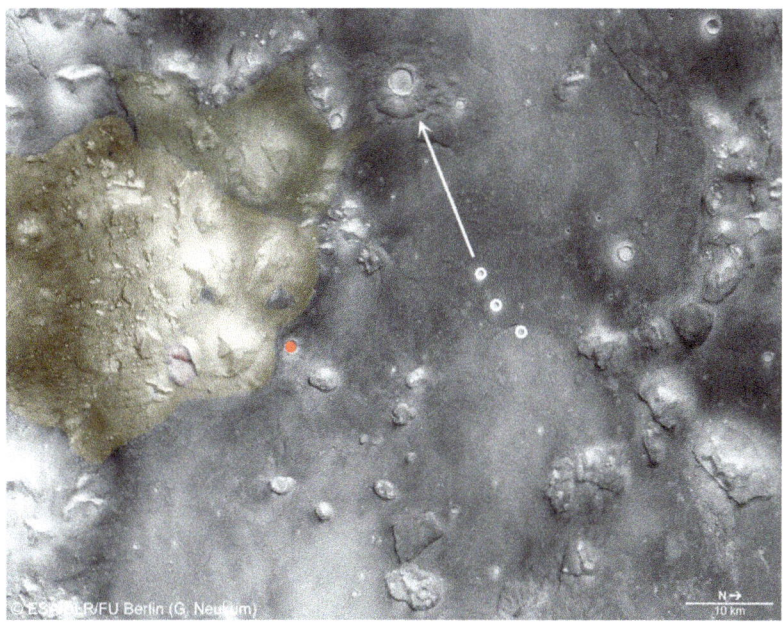

3.4 Three "Orion" craters point to larger crater.

SIRIUS THE DOG STAR

The Greater Orion family of constellations includes Orion, Canis Major, Canis Minor, Lepus, and Monoceros. In Greek mythology, this group of constellations consists of the hunter, Orion, and his two dogs, Canis Major and Canis Minor, chasing the hare, Lepus. Figure 3.5 is a star map showing the Upper Canis Major constellation. The circle size corresponds to the star's intensity.

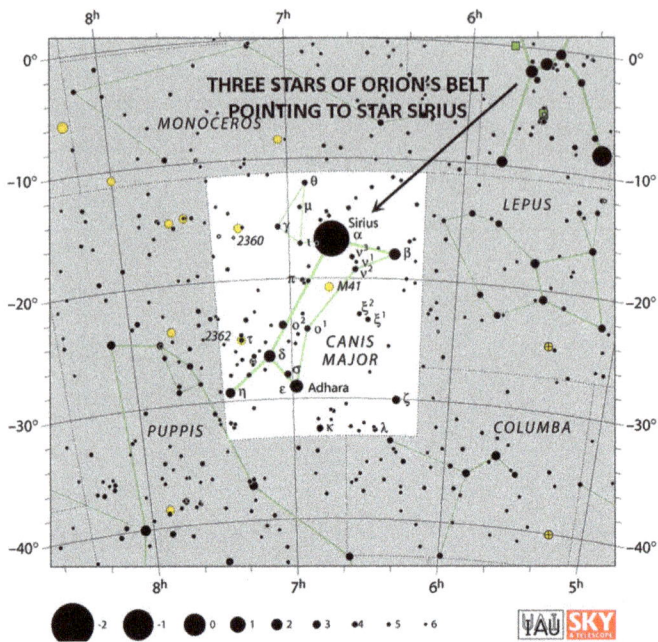

3.5 Star Map of The Constellation of Greater Orion.
Courtesy, International Astronomy Union.

Looking at this star map, one can see that the three stars of Orion's belt point to the star Sirius just as in the Cydonia alignment. If one lays a crop of the star map showing Sirius and Mirzam overtop of the two Cydonia craters, it forms a perfect match in their relative position and intensity.

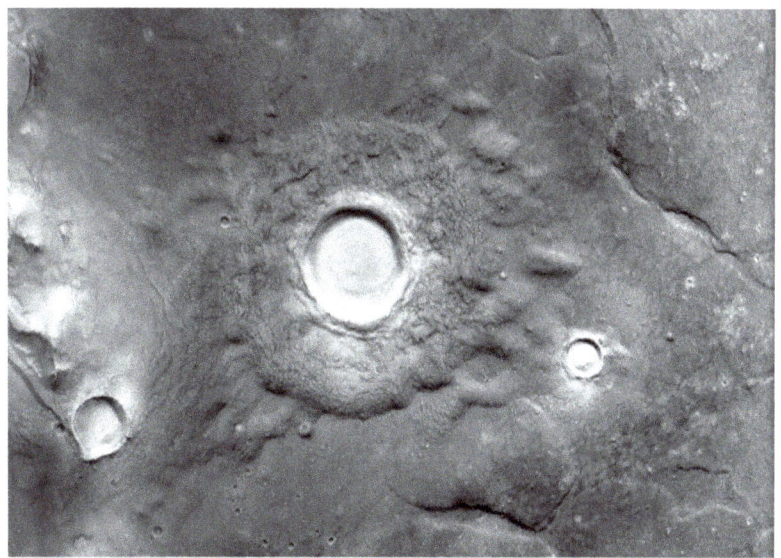

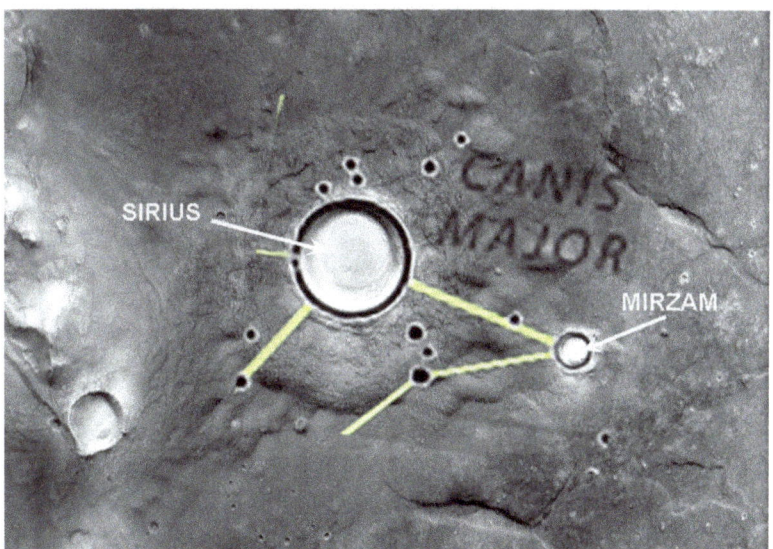

3.6 Cydonia Mars craters overlain with Sirius star map.

If the Cydonia image in figure 5.6 is rotated 23 degrees to the left, it brings a half face of another lion into position. When this half-face is mirrored along an axis of demarcation created from a prominent mound and ridge, the full face of a lion in its guise as a sun appears. The crater representing the star Mirzam becomes the Sun-lion's ear flares (Figure 3.8)

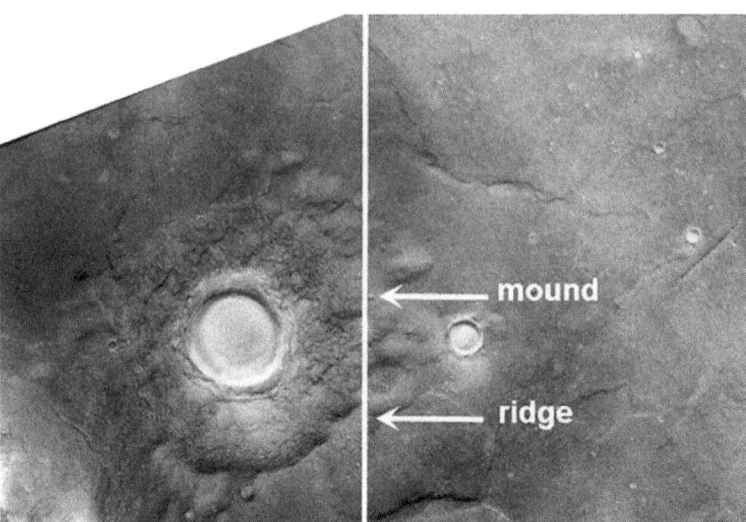

3.7 Image demarcation line defined by mound and ridge.

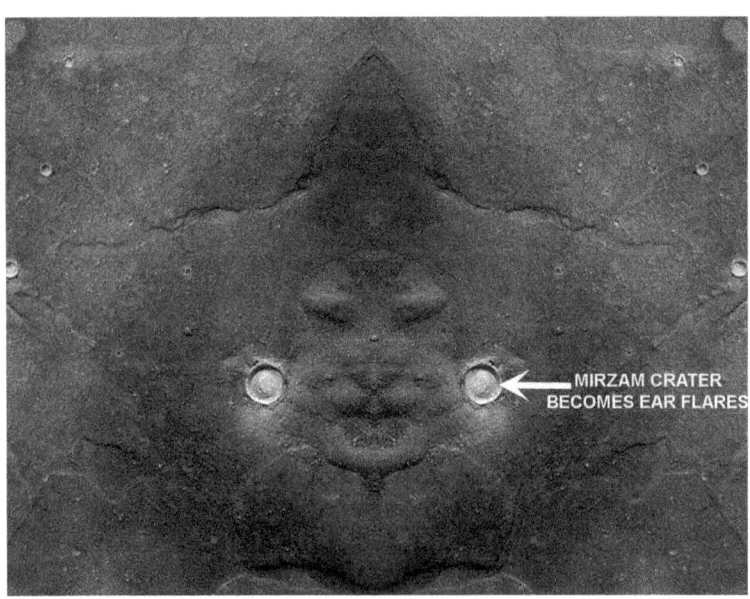

3.8 Mirrored Sirius Sun-Lion with ear flares.

In Mesoamerican culture, Jade ear flares (also called ear spools) represented the Sun, portals to the Otherworld, and conduits for spiritual energy.[1] They can be found in a multitude of Central and South American art work.

3.9 Aztec Sun Stone showing ear flares.

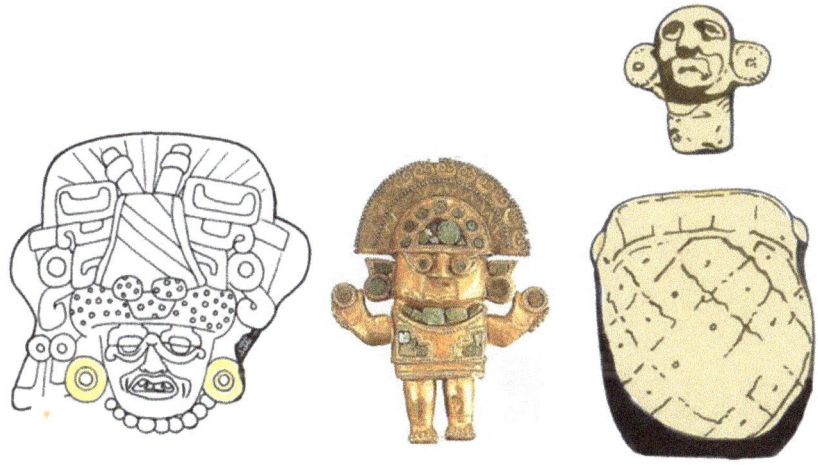

3.10 Left: Aztec god Cocijo; center: Inca Sun god;
Right: Maya Sky Bearer god Pawatun;
all wearing ear flares or ear spools.

Returning to the large crater in Cydonia, one can see it has a circular ejecta apron and a circular center (Figure 3.11). This matches the image of the circumpunct star, which is a sun symbol and as well as a symbol of the Egyptian god Ra.

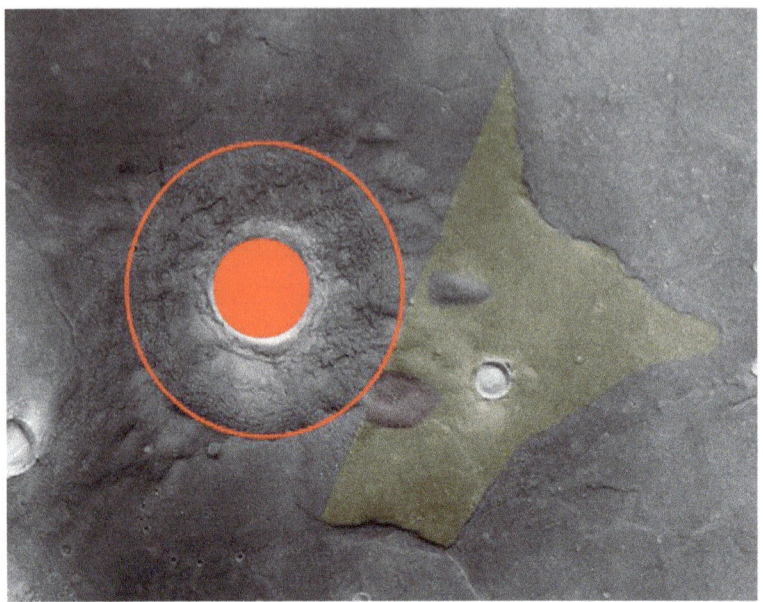

3.11 The half face of sun-lion with circumpunct symbol. Color added by author.

THE STAR AND CRESCENT

To the north of the crowned lion and the Sirius Sun-lion are numerous mesas, which together form a crescent shape.[2] In front of this crescent is another crater. When a star map of the upper portion of the Orion constellation is overlaid, the crater matches the star Betelgeuse with its size on the intensity scale. As well, a smaller crater matches another Orion star believed to be Meissa, while a mountain peak at the bottom of the crescent matches the position of the star Bellatrix. (Figure 3.13)

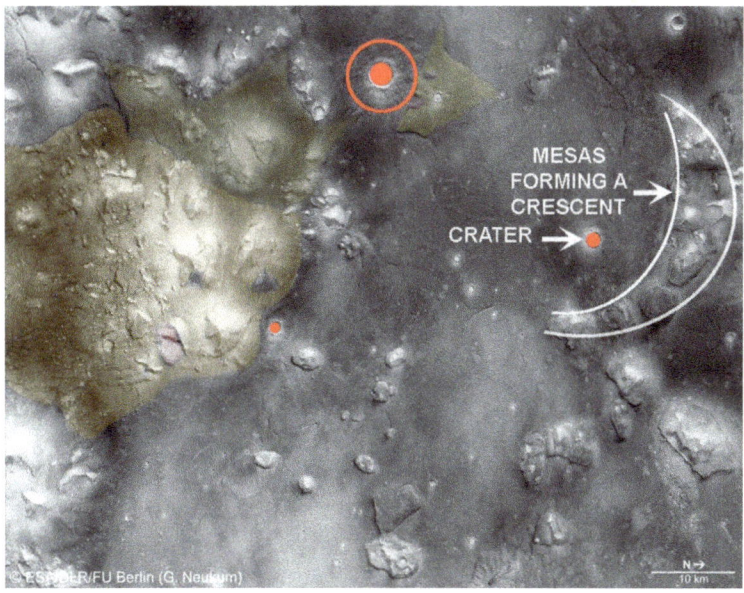

3.12 Area of Cydonia showing crescent formation with crater in front.

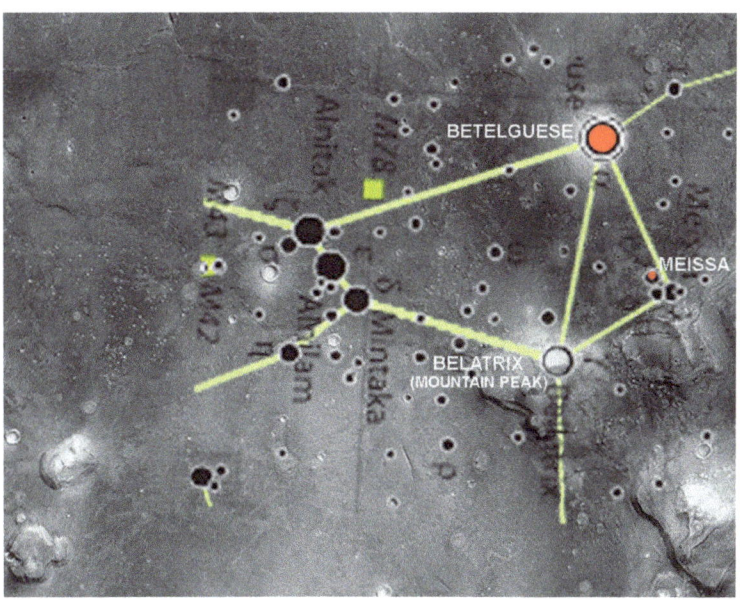

3.13 Star map of Orion overlies Cydonia craters and mountain peak.

By joining the three star-craters, from figures 3.2, 3.6, and 3.14, they form an isosceles triangle. (Figure 3.14).

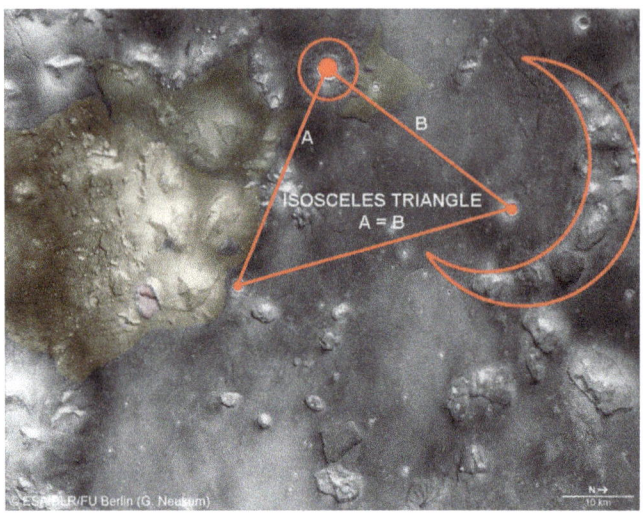

3.14 The three "star craters" form an isosceles triangle.

Presented in figure 3.16 is a star map from the International Astronomy Union (IAU) along with its mirror image. If you place the star map over the Cydonia landscape, the three stars of Sirius, Betelgeuse, and a star in the constellation Lepus, Epsilon Leporis, overlie the three star-craters. Sirius' companion star, Mirzam, points directly at the star Epsilon Leporis. (Figure 3.17)

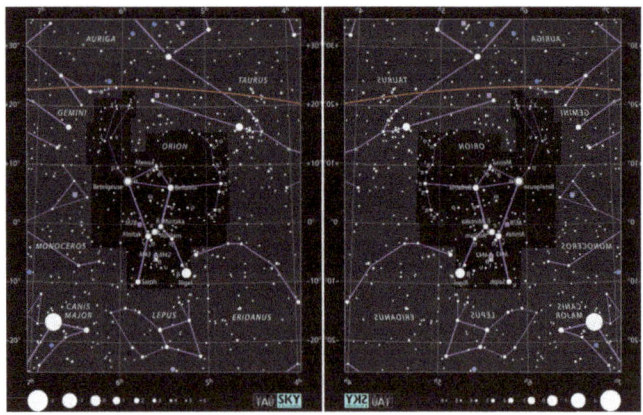

3.15 Left: Star Map showing Greater Orion with
Canis Major and Lepus constellations.
Courtesy International Astronomy Union.
Right: Mirror image of Star Map.

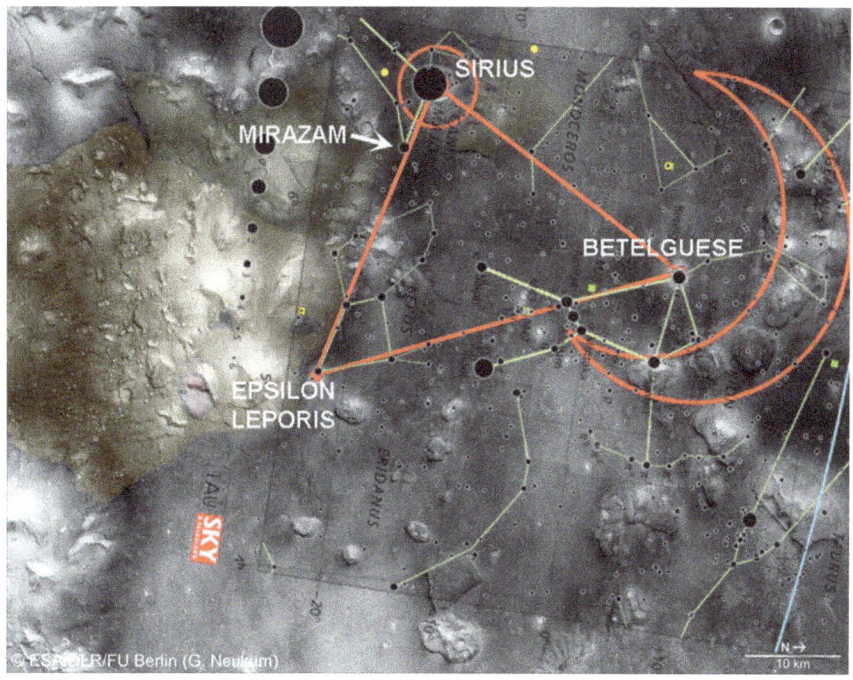

3.16 Star Map shows Sirius, Betelguese and Epsilon Leporis overlying the three Cydonia craters with Mirazam pointing directly to Epsilon Leporis.

When one replaces the star map with its mirror image, the stars Betelgeuse and Epsilon Leporis switch positions, and the star Mirzam lies very near its position on the Cydonia landscape. Perhaps the small difference is due to the passage of time.

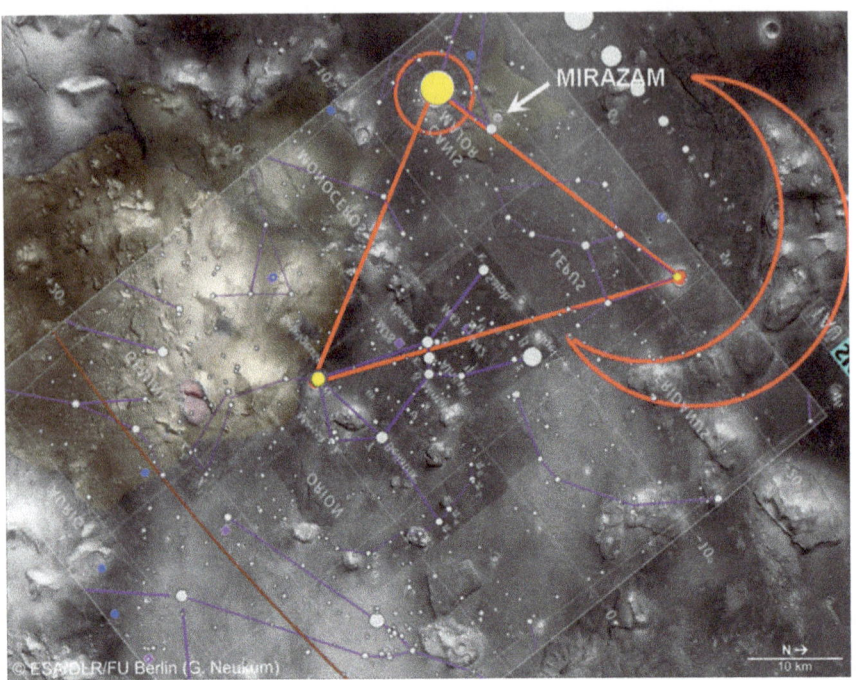

3.17 Mirrored Star Map shows Betelguese and Epsilon Leporis switching positions, and now the star Mirzam lies very near its position on the Cydonia landscape.

From the ancient archaeological site at Mount Nemrut in Turkey is the carving of a lion believed to be over 2000 years old (Figure 3.18). The lion is adorned with stars, including the three Orion belt stars above its back while a crescent hangs below its neck. Notice the extended tongue just as the Cydonia lion. This ancient relic appears to be relaying similar information to what we are finding in Cydonia, Mars.

NEMRUD-DAGH WESTTERRASSE
HOROSKOP DES ANTIOCHOS

3.18 Believed to be 2000 years old Carved **relief** depicting the Lion
Horoscope found at Nemrut Dagi in Leo with Pisces. Image cast
by Carl Humann, 1883. Photographer unknown, public domain.
Note correlations to Cydonia; the lion's tongue extended,
the star on its nose, the crescent with the star in front,
and the three stars of Orion's belt along the back.
Tongue and crescent color added by author.

THE DOG AND THE NURSERY

Sirius, the largest and brightest star of Canis Major, is known as the Dog
Star, and if one looks just below the large crater, in the ESA photo from
figure 3.1, there is the profile of a dog. As is shown in the detail image,
figure 3.20, the dog is suckling at a breast.

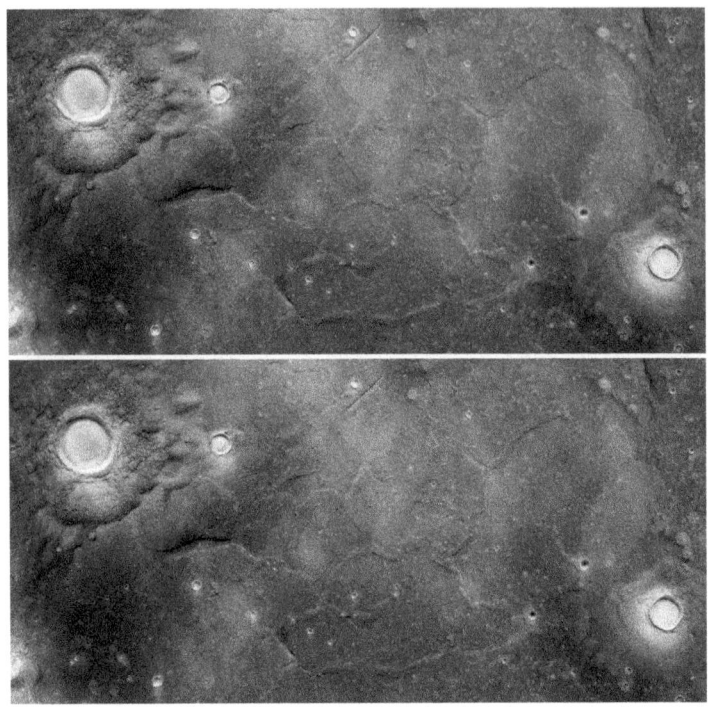

3.19 Dog profile. Detail from ESA 300-230906-3253-6-nd1. Color added by author on bottom image.

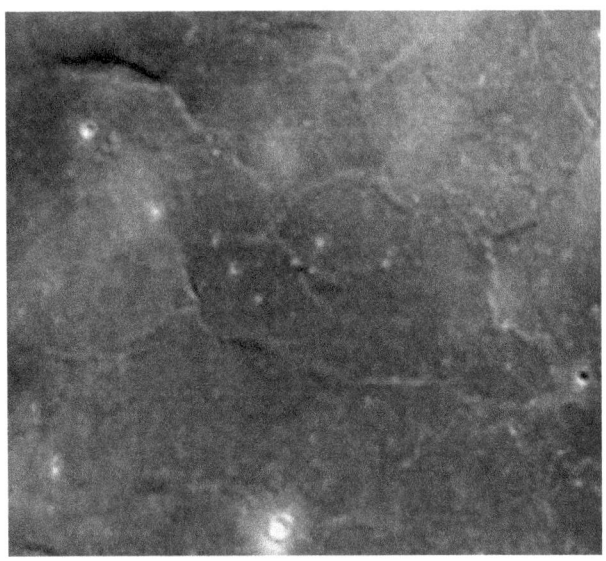

3.20 Detail from 3.19. Suckling dog. Color added by author.

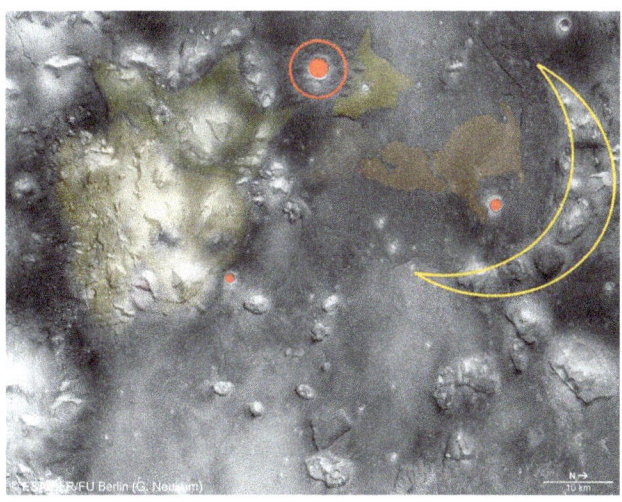

3.21 A portion of the Cydonia, Mars
landscape with highlighted images.

If one extends the line of demarcation used to create the Sirius Sun-lion in figure 3.7, another fascinating scene is created. When the image is mirrored, two dogs suckle at the anthropomorphized Sun-lion. Is it surprising then to learn that astronomers consider the constellation of Orion a nursery for new stars?

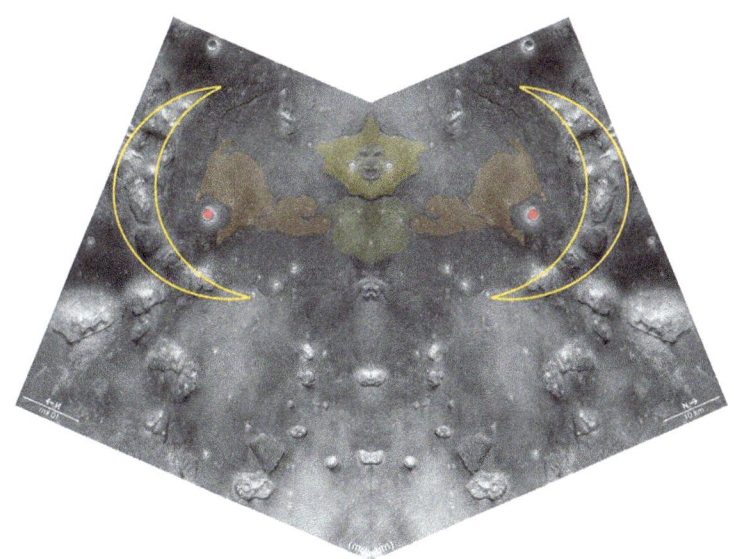

3.22 Mirrored Sun-Lion with dogs suckling.

Consistent with the connection to Mesoamerican iconography in these Mars designs comes a pre-Columbian sculpture from Colima, Mexico over 1500 years old, of a woman nursing a young dog as she looks skyward (Figure 3.23).

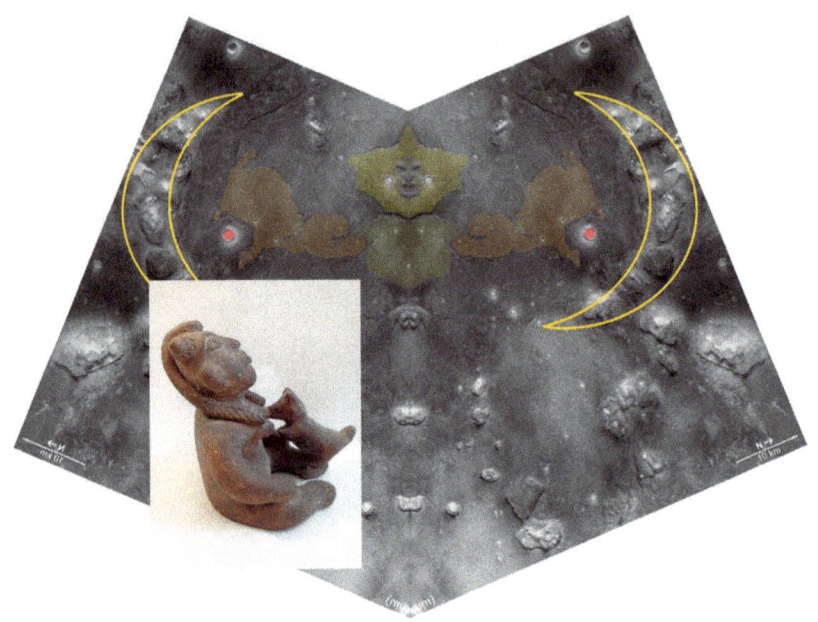

3.23 Mirrored Sirius Star Lion with suckling dogs and Colima statue with woman suckling a young dog as she looks skyward.

From the other side of the world, dated to around 800 B.C.E., is an ancient Sumerian lioness-headed deity, Lamashtu, pictured with two dogs suckling at her breasts (Figure 3.24). Once again we are seeing duality presented. You see, the Sumerian deity Lamashtu is said to be the most terrible of all female demons. Daughter of the sky god Anu, she slew children and drank the blood of men and ate their flesh.[3] So here we have a depiction on Mars presenting Sirius as a mother giving birth and the identical imagery in an Earthly representation of the demon Lamashtu, which ancient lore says devours children. Is the Sumerian demon Lamashtu a symbolic indication that, although Sirius may have given birth to new stars, she may eventually cause their demise?

3.24 800 BCE, Neo-Assyrian lion-headed deity
Lamashtu with two dogs suckling.
Amulet, Copyright British Museum Images 00032627001

Another image pertaining to birth can be produced by rotating the half image of the Sirius Star-lion, from figure 3.7, 33 degrees counter-clockwise (Figure 3.25). Mirroring this rotation along a line formed from a half-face and a few small mounds creates a picture of a person presenting the star Sirius and its ejecta apron as a vagina. Two small dogs, appearing to emerge from the edge of the Mirzam star crater, nuzzle against the person's arms (Figure 3.26 & 3.27). They also each have a paw upon the head of a pup. (Figure 3.29). In addition, a fascinating similarity exists in comparing the shape of the entire image with that of a uterus (Figure 3.28)!

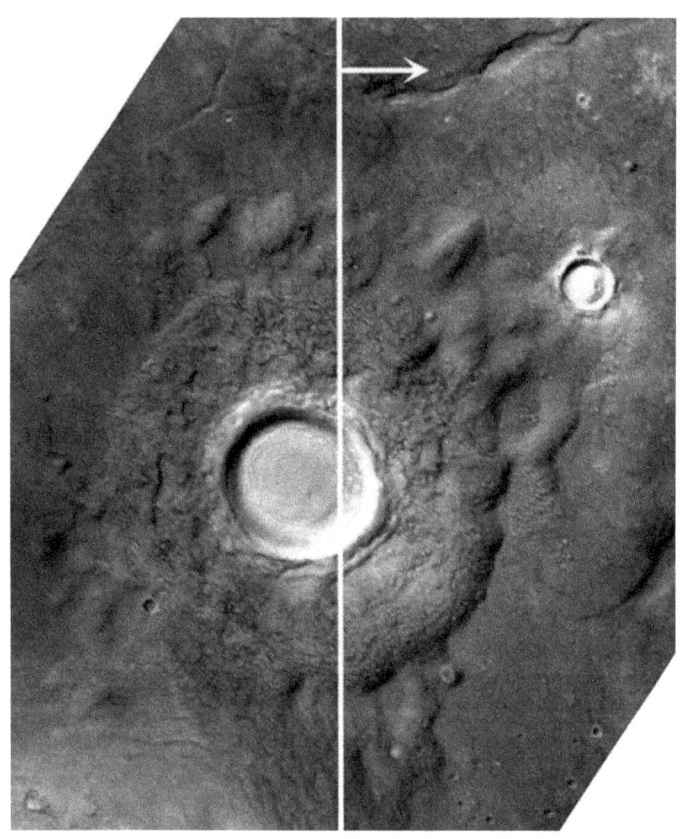

3.25 33 degree counter-clockwise rotation of figure 3.7 showing demarcation line for figure 3.26.

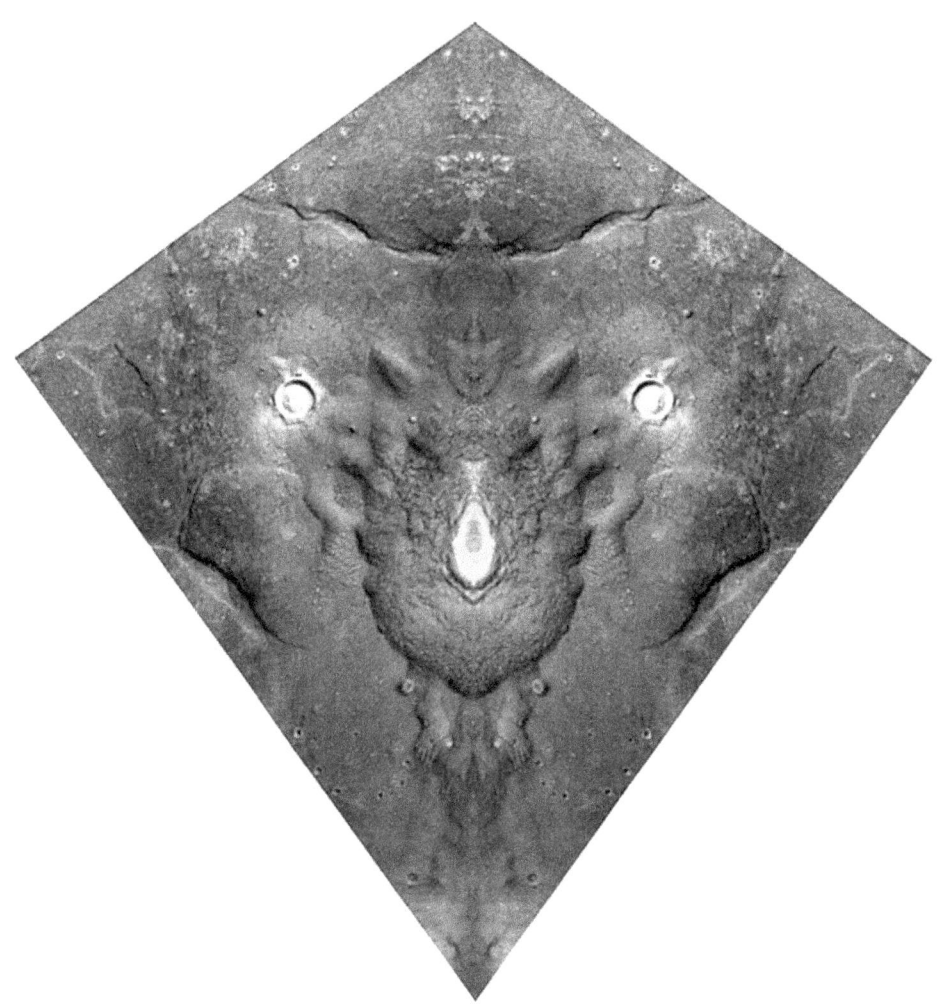

3.26 Mirrored half image from 3.25 showing the star Sirius and ejecta apron as a vagina.

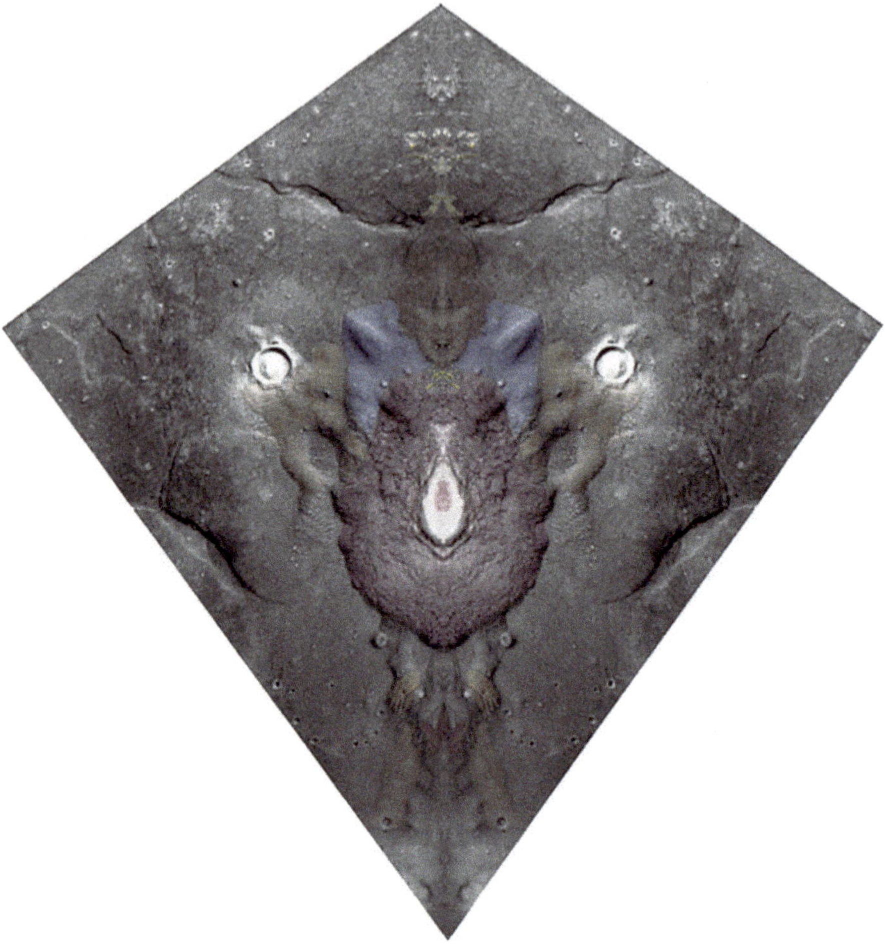

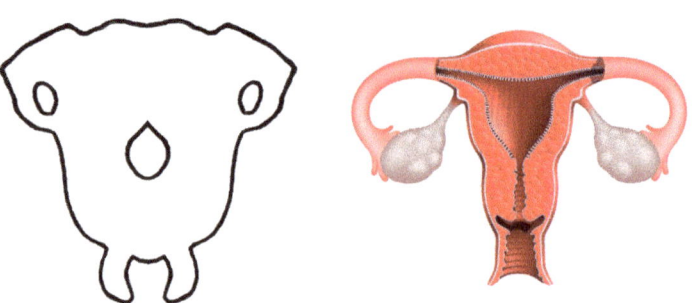

3.27 Color version of 3.26. Mirrored half image from 3.25 showing the star Sirius and ejecta apron as a vagina. Color added by author.

3.28 Left: The outline of figure 3.26; Right: a human uterus.

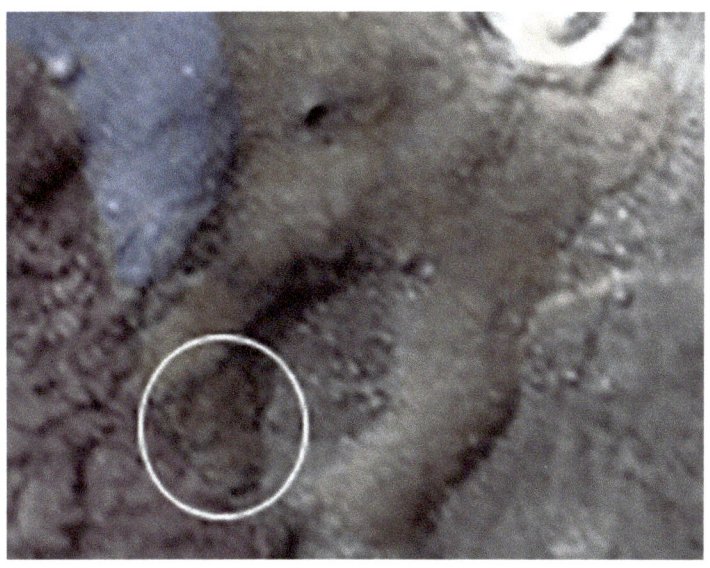

3.29 Detail from figure 3.27. Dog holds its paw on the head of a pup.

THE SUN LION CHARIOT

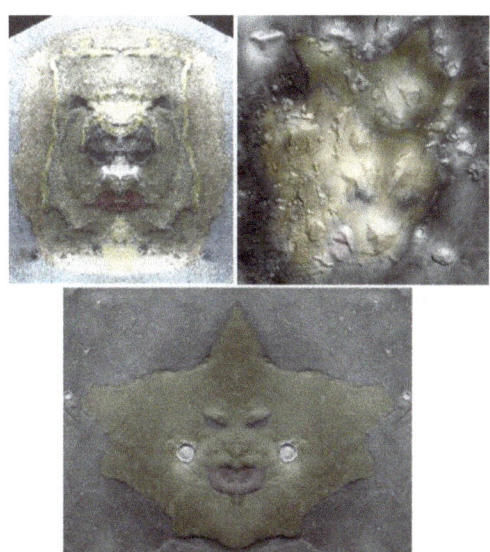

3.30 Three lions in Cydonia.
Left: Mirrored Right side of 'The Face on Mars'

Right: Profile of Crowned Lion; Bottom: Mirrored Sirius Star Lion

Up to this point, we have seen three representations of lions, the feline half of the 'Face on Mars, the gigantic lion with the three-point crown, and the half-face of a lion on the Sirius crater apron. A fourth lion, in the crop of an ESA image, continues the repetitive symbolism.

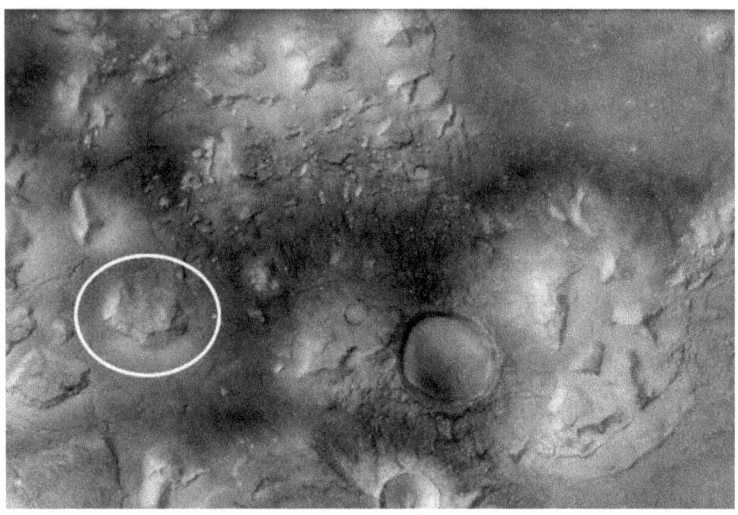

3.31 Lion-faced mesa, cropped portion of ESA H3253 Courtesy European Space Agency

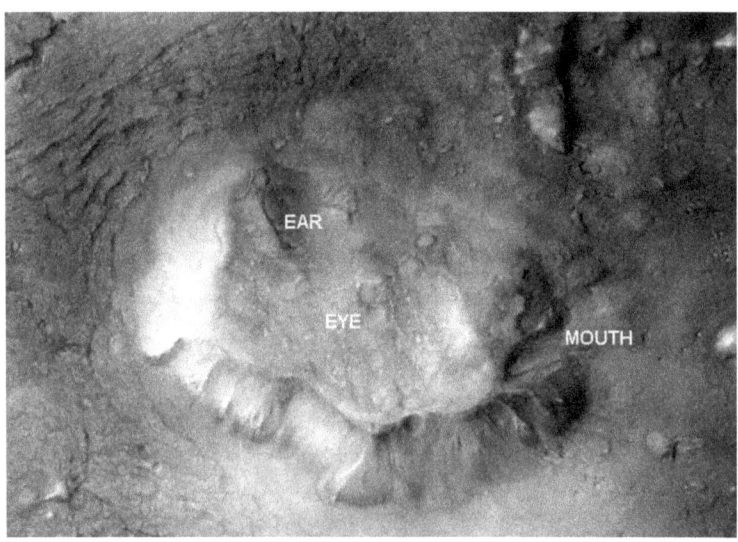

3.32 Detail, Lion-Faced mesa.

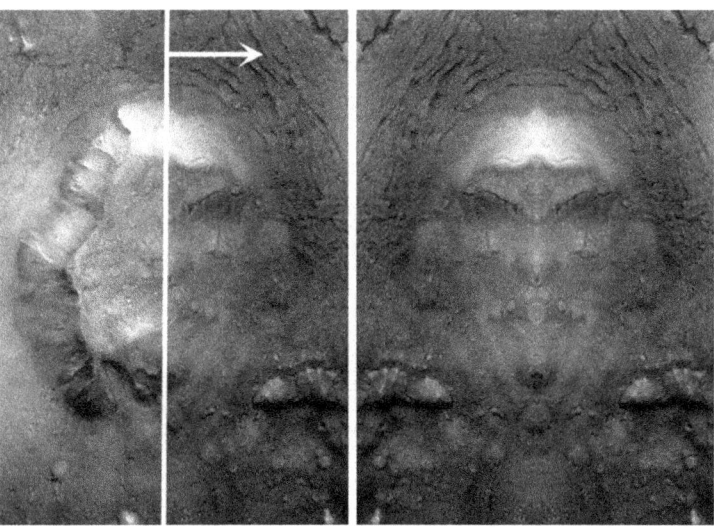

3.33 Left: Lion's face from image 3.32 rotated 90 degrees; Right: Image mirrored.

By rotating the lion-faced mesa 90 degrees clockwise, the half-face of a man can be seen on the right side By mirroring the entire photograph along this line, a person, wearing a moustache and chin-beard appears at the controls of, what I call, a Sun-Lion Chariot (Figure 3.34, 3.35)! To either side, emerging from the lions' manes, are twin human heads with a look of shock on their faces. The chariot is being pulled by 3 lions, the Sun-Lion wearing the three point crown from figure 3.2, his mirrored image being the second lion and a third, bottom center, looking up at the driver. An ancient Greek coin depicts a similar scene where the Sun god Apollo, sun rays emanating from behind his head, stands upon a chariot pulled by three lions (Figure 3.36).

3.34 Sun-Lion Chariot. Portion of ESA H3253
mirrored along half face from image 3.33.

3.35 Mirrored Sun-Lion Chariot Driver. Detail
of 3.34. Color added by author.

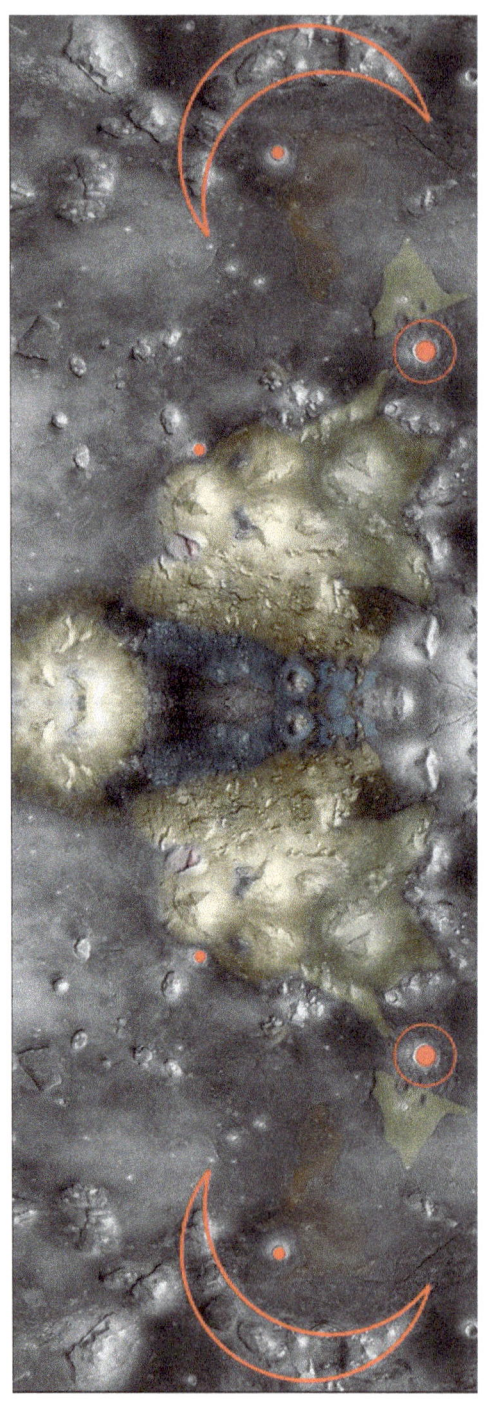

3.36 Mirrored Cydonia, Mars with highlighted features.

3.37 Ancient Greek coin depicting the god Apollo and a chariot with 3 lions.

FOOTNOTES:

1. Marilyn A. Masson and David A. Friedel, *Ancient Maya Political Economies*, (Walnut Creek, Calf., Altamira, 2002) 63. Also; https://onetribe.net/blogs/content/78343681-status-symbolism-and-spirit-of-the-mayan-ear-flare
2. From research over the last number of years, the Cydonia Institute has documented a large amount of Mayan iconography associated with these mesas which is awaiting publication.
3. https://www.britannica.com/topic/Lamashtu

CHAPTER IV

THE MOON AND THE MOON GODDESS

"It's easier to explain the Moon's non-existence than it is its existence." - Robin Brett former NASA scientist

THE MOON GODDESS ON MARS

The D&M pyramid on Mars is a structure that became well known in the early investigation of the Viking Orbiter images in the late 1970's and early 1980's. It got its name from Vincent Dipietro and Greg Molanaar, the first researchers to publish data on the structure. In the 1976 Viking Orbiter images, the structure looked like a five-sided pyramid with highly significant geometrical angles (Figure 4.1)

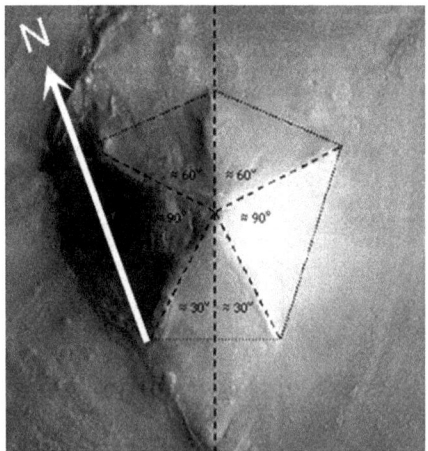

4.1 D&M Pyramid; 1976 Viking Orbiter (70A13).
Courtesy Dr. Mark J. Carlotto.

However, the structure looked a bit different in images taken in 1998 with the new Mars Orbital camera onboard the Mars Global Surveyor spacecraft (Figure 4.2). Some of the geometric angles were still there, but the appearance of two of the buttresses was not as it seemed from further out in space. I believe this is an intentional design, part of the genius involved in the creation of these structures. From different perspectives, their appearance often changes, as they present a different yet meaningful form.

What appeared as a symmetrical five-sided pyramid from 1000 miles up transformed into a complex mix of shapes, lines, and curves when viewed from 250 miles up. Most researchers, who believed the structure to be an artificial, five-sided pyramid, judged this complexity of lines, curves, and shapes to be a sign of imperfection, attributable to erosion or war, perhaps. Though the conjecture of war may be true in part, I take another position. I believe this structure, along with others on the surface of Mars, was designed with information layered within viewing perspectives.

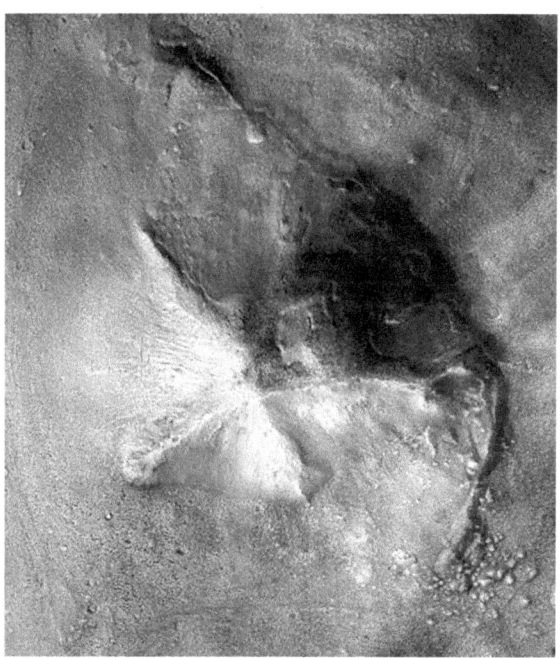

4.2 D&M Pyramid, A section of composite image (2003). Courtesy Keith Laney and Jeff Williams. R06-00469, R07-00422, R11-02437 Mars Orbital Camera on Board the Mars Global Surveyor Spacecraft.

So what appears as a highly geometric, five-sided pyramid, from further out in space, contains the profile face of a young woman wearing a canine headdress with the half image of another woman in front of her face (Figure 4.3). The upper portion of the gown is on one side of the ridge while the lower half continues draped over into the next section of the structure.

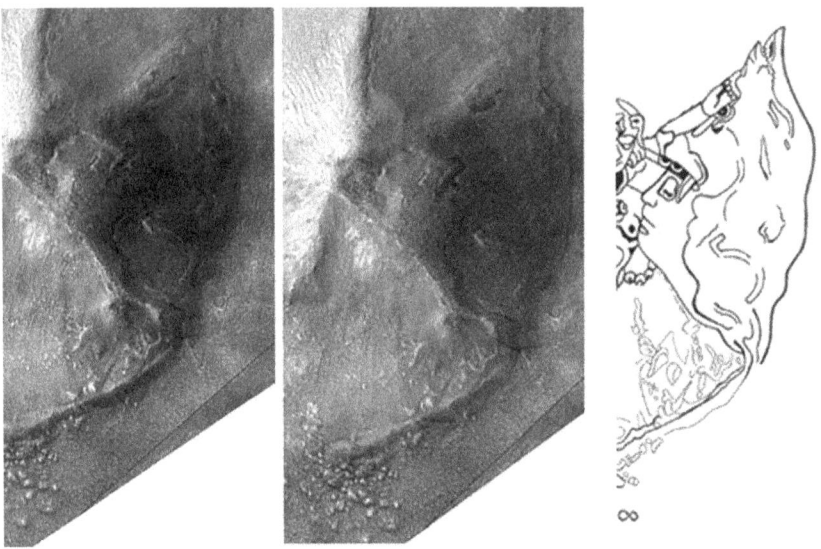

4.3 Left: Eastern portion of The D&M Pyramid;
Center: false color highlights;
Right: Drawing by George J. Haas

Many presentations of the Moon goddess in Maya art depict her with a canine companion that has a small, tattered ear and long back-curving tongue, just as is seen in the Mars image. The young woman appears to have her eye closed; however, she may actually be wearing an eye patch. In Maya mythology, the young goddess had one of her eyes torn out by the Sun as a result of the early inhabitants of Earth complaining to the gods that the moon was so bright they could not sleep at night.

4.4 Left: Xolotl – Maya, drawing by George J.
Haas. (Image source: Codex Borgia, p.65)
Right: Moon Goddess, Mars, cropped from figure 4.2.

When mirrored along the axis of the half-woman, the (Full) Moon goddess is created. Standing with her eyes lowered and her head slightly bowed, she presents an innocent and humble demeanor (Figure 4.5). She wears a beautifully embroidered gown that drapes over one of the strucutre's ridges and continues down the other side. Her breasts are revealed, and there is a bracket of patterned stones on the ground below. She is flanked by two other Moon goddesses kissing her breasts. My colleague, George Haas, and I believe they represent the waning and waxing moons.

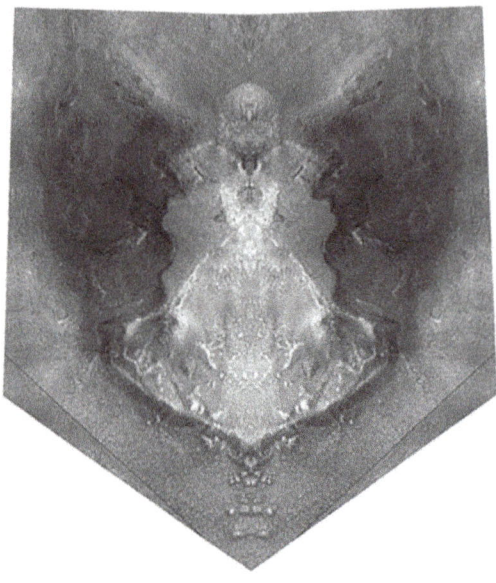

4.5 Mirrored full Moon Goddess at center with waning and waxing Moon Goddesses on either side, kissing her breasts.

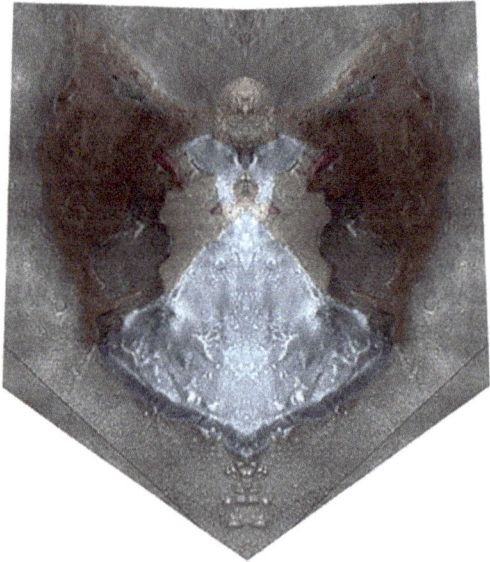

4.6 Mirrored full Moon Goddess at center with waning and waxing Moon Goddess on either side kissing her breasts. False color added by author.

A Maya version of the Moon goddess, from the Madrid Codex, is depicted with bare breasts and outstretched arms (Figure 4.7). She also wears a flowing skirt and a bundled serpent headdress. Below her dress we find four dots, while water flows from her breasts. It is said this flowing water denotes the moon's effect on tides and the rainy seasons. The flow from her skirt likely represents the menstruation cycle, which closely matches the 27.3-day Moon cycle and thus gave the Moon goddesses of ancient cultures the association with fertility. The Maya blood scroll glyph, seen in figure 4.7, has 4 dots along the bottom.

4.7 Left: Maya Moon Goddess. Drawing by George J. Haas. Image source: Madrid Codex.

Above the mirrored Mars Moon goddess is a round, crater-like depression, representing the Moon. A highly reflective, crescent-shaped ridge represents the waning and waxing Moons.

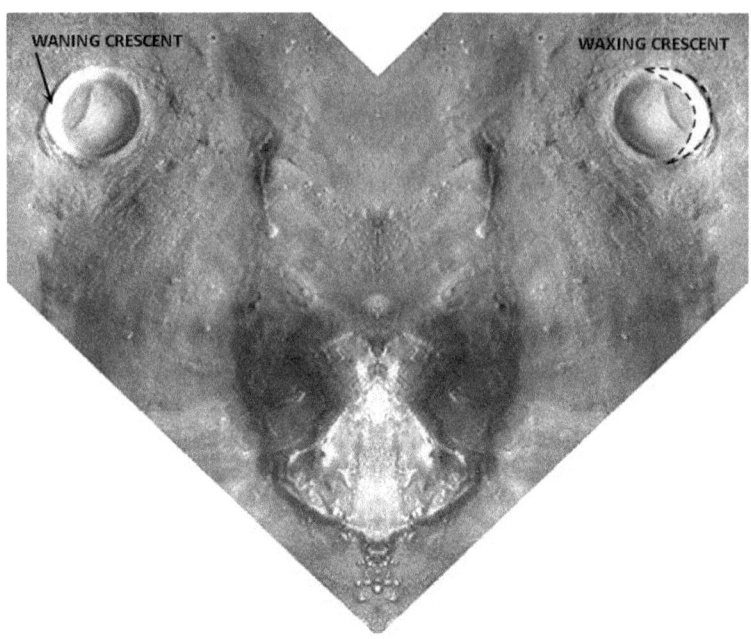

4.8 Moon Goddess with waning and waxing
Moon crescents above her.

In Classic Maya art, the Moon goddess is often shown accompanied by a rabbit known as One Rabbit, and framed by the crescent of the waxing Moon, her most important identifying attribute. At the tip of the D&M structure, between the Moon goddess and the crescent Moon, is the profile of a rabbit or hare.

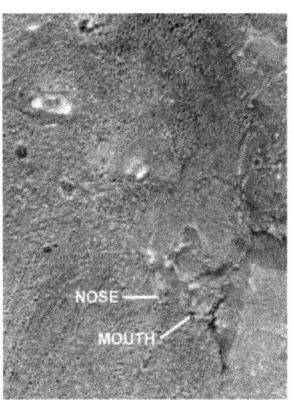 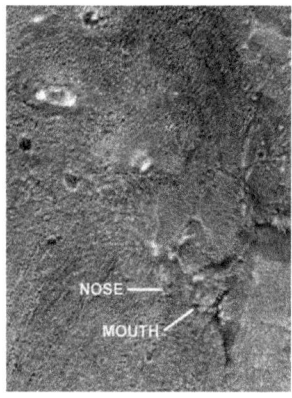

4.9 Left: Rabbit (Hare) profile; Right: Rabbit (Hare) profile, false color.

The Moon goddess also had a tattered-eared canine companion called Xolotl, (pronounced: She-ol-otl) who would lead the Moon goddess to the Underworld at night. Immediately below the D & M where the Moon goddess is located, we find another half image built into a mesa (Figure 4.10).

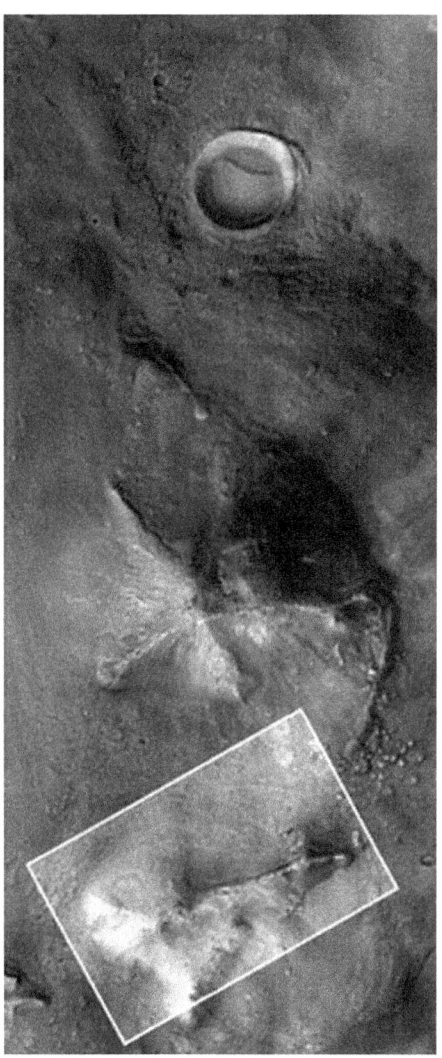

4.10 Location of half image of Xolotl
Composite image (2003) R06-00469, R07-00422, R11-02437
Mars Orbital Camera on Board the Mars Global Surveyor Spacecraft.
Courtesy Keith Laney and Jeff Williams.

When duplicated along the axis determined by the structural peak at the nose and a mound in the center of the forehead, we have the Moon goddess' companion, Xolotl (Figure 4.11). If we look closely at the tip of its ear, we can see its tattered nature.

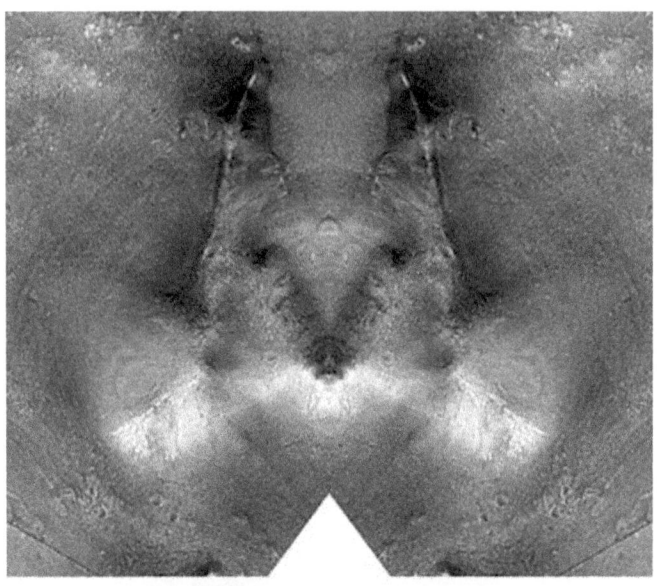

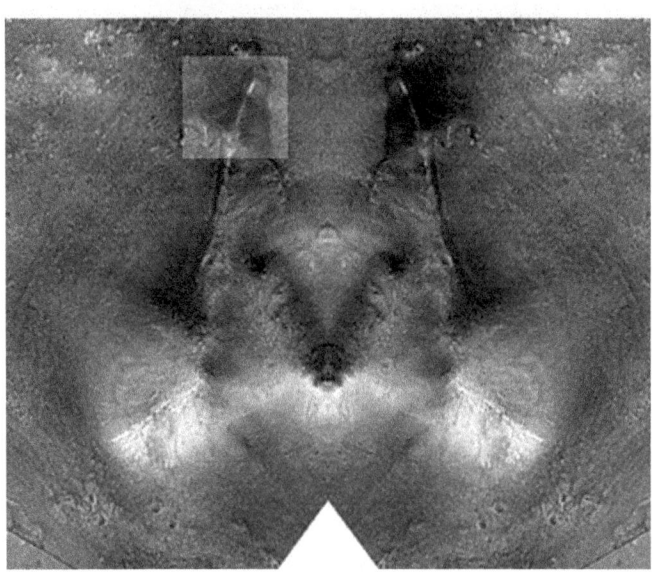

4.11 Top: Xolotl, Mirrored Half Image;
Below: False color, notice the 'tattered ear' highlighted by author.

THE DARK MOON GODDESS

With the discoveries of the profiles and half faces on Mars, I started to believe the face of 'The Man in the Moon' was an intentional creation; however, I never got around to investigating the Moon for possible art similar to Mars until I was preparing this book.

The Moon goddess can be found in the mythology of most ancient cultures, and my colleague, George Haas, and I had found a lot of symbolism related to the Moon goddess on Mars. So it would seem natural to expect some similar imagery on the Moon as well.

4.12 Super Moon, January 2019 through telescope.

A thumbnail picture of a January 2019 super Moon on my desktop computer kept grabbing my eye, so I finally brought it into my photo software and began to search for the Moon goddess. One of the things I noticed first was the dark area on the left side appeared to be the profile of an animal such as a bear or a lioness. The dark areas of the Moon are from flows of basaltic rock that cover the low areas. The bright areas are higher elevations reflecting the sunlight. They are believed to be the result of meteor impacts and possibly ancient volcanic remnants. Experience from the Mars research told me that if there is a profile, there is likely to be a half image beside it so that when mirrored, profile frames the whole picture.

There seemed to be a guiding hand in many of our Mars discoveries. Whose hand, I don't know, but many of our discoveries have had the demarcation line for mirroring right down the center of the spacecraft's photographic swath. This is not a spacecraft image; it is from a telescope. As only one side of the Moon is visible from Earth, however, it presents the same situation.

While determining the center of the photograph, a half-face started to come into focus. Using the edge of the half-face as a guide, I continued the demarcation line down the entire image. Upon completing the mirroring process, a face wearing an elaborate crown or headdress appeared. At first, I was ecstatic at what I was seeing; a Moon goddess with a royal headdress and framed by lionesses. However, as I looked closer, expecting to see a beautiful Moon goddess,

4.13 Moon Goddess demarcation line down the center of the picture. The half image is on the left side.

I was shocked! Something was wrong; the Moon goddess had been defaced! Where was the pretty and humble Moon goddess witnessed on Mars?

I played with the demarcation line a few times, but it did not change the fact that this was not a pretty Moon goddess. How could this be? Everything else had fallen into place. I closed my computer in disgust. I would try again another day. I was still busy putting this book together and had other priorities for the time being.

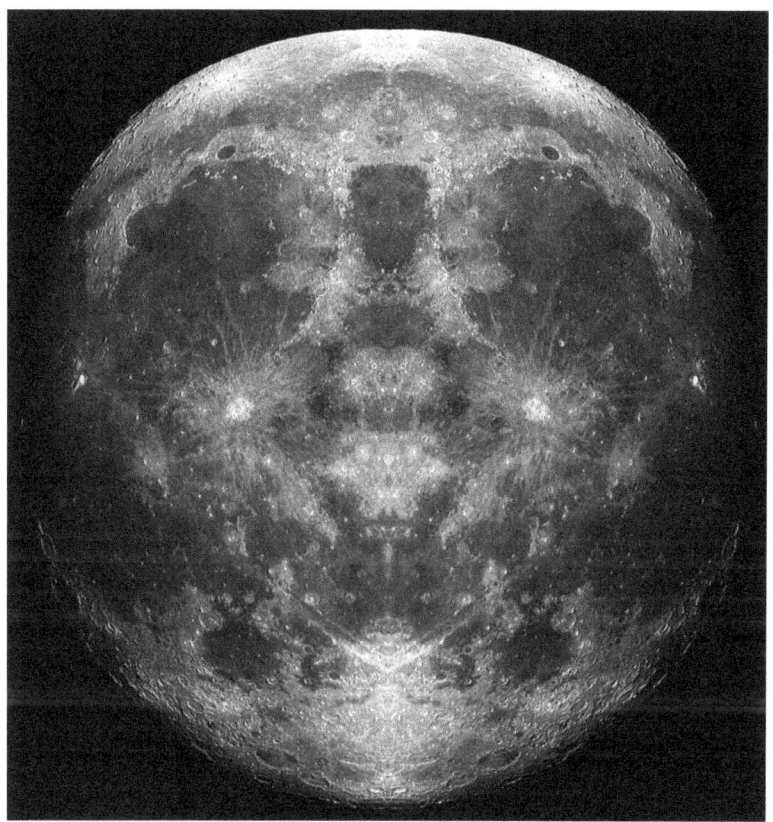

4.14 Mirrored image of Dark Moon Goddess

The Moon goddess image haunted me for days until I finally had an "Ah hah" moment! If that face was an intended image, which I was sure it was, then it was supposed to look that way, and I needed to find out why. I brought up the image once again and looked closely; what I saw was a bird face. Together with abundant bird iconography having been discovered on

Mars, birds are an important element of Maya and Egyptian mythologies; I needed to do more research. From the website Transcendence Works, Moon Goddess Compendium:

> *"Ancient Mesopotamia spawned continuously evolving pantheons over its rich history. Ishtar, or Inanna, was the Babylonian goddess of the moon, as well as of love, fertility, passion, sexuality. She was symbolized with the moon, stars, lions and doves."* [1]

So, in Ancient Mesopotamia, we see the Moon goddess associated with lions, which we see framing her mirrored image on the Moon, and doves, which are birds. So far, so good. The website looks at the Moon goddess from other cultural points of view, as well.

> *"Lilith, widely called goddess of the dark moon, often represented as a demon goddess, a purely carnal, sexual being. The widely accepted myth of Lilith as a demon represents Western patriarchal society's categorical rejection of female sexuality and individual autonomy. Her threat is her independence and refusal to obey men, unlike the more obedient Eve in the Biblical Garden of Eden, and her unwillingness to be controlled created discomfort at important points in history. Patriarchal rewritings of ancient myth occurred close to 3500 years ago, and tell us that she was condemned to lie with demons and monsters, making her own children beasts. Prior to this cultural shift, within the pantheon of ancient Sumerian gods and goddesses, Lilith was known as Lilitu, a great winged bird goddess. She was recorded in Sumerian myth as a young handmaiden to the Queen of Heaven and Earth, Inanna. Lilith brought the men from the streets into Inanna's temple to engage in sacred sexual rites. Hebrew religion began to value women either as virgins or property of their husbands, so this old belief of the Sumerians needed to be changed, and the authorities found a way to change it into a tool of repression—exactly that*

which the Judaic Hebrew texts state that Lilith vehemently opposed." [2]

So Lilith was known as the goddess of the Dark Moon. The image in figure 4.15 definitely has a dark complexion. She was also known as "a great winged bird goddess", so that helps explain the bird's beak. It also says that she was condemned to lie with demons. In figure 4.16, one can see what can be described as a demon's head and face just below her neck area. The demon has a dark head and a white, highly ornamented face with piercing bright eyes. From the website Thought Co., Gods and Goddess of the Maya:

> "The Maya Moon goddess *Ix Chel, or goddess I, is a frequently clawed goddess who wears a serpent as a headdress. Ix Chel is sometimes illustrated as a young woman and sometimes as an old one. Sometimes she is portrayed as a man, and at other times she has both male and female characteristics. Some scholars argue that Ix Chel is the same deity as Chac Chel; the two are simply different aspects of the same goddess. There is even some evidence that Ix Chel is not this goddess's name, but whatever her name was, Goddess I is the goddess of the moon, childbirth, fertility, pregnancy, and weaving, and she is often illustrated wearing a lunar crescent, a rabbit and a beak-like nose. According to colonial records, there were Maya shrines dedicated to her on Cozumel Island."* [3]

So, there it is! The Moon goddess was associated with lions and doves in ancient Babylon. In ancient Sumer, the Moon goddess, Lilitu, was known as the "Great Winged Bird Goddess". In Hebrew mythology she was Lilith, and condemned to lie with demons! And in Mesoamerica, she was commonly depicted with a serpent headdress and a bird's beak! It is all there. On the Moon is the depiction of a Moon goddess, hidden in the form of a half image. When mirrored, the resulting portrait expresses the symbolism of her mythology, from different cultures, over a large time frame.

This Dark Moon goddess seems to be a different aspect of the Moon goddess we see depicted on Mars with her canine companion Xolotl. Would he also be present with her on the Moon? Well, when one mirrors the opposite side of the demarcation line for the Moon goddess from figure 6.13, one is again presented with the canine companion, Xolotl (Figure 4.16 & 4.17.

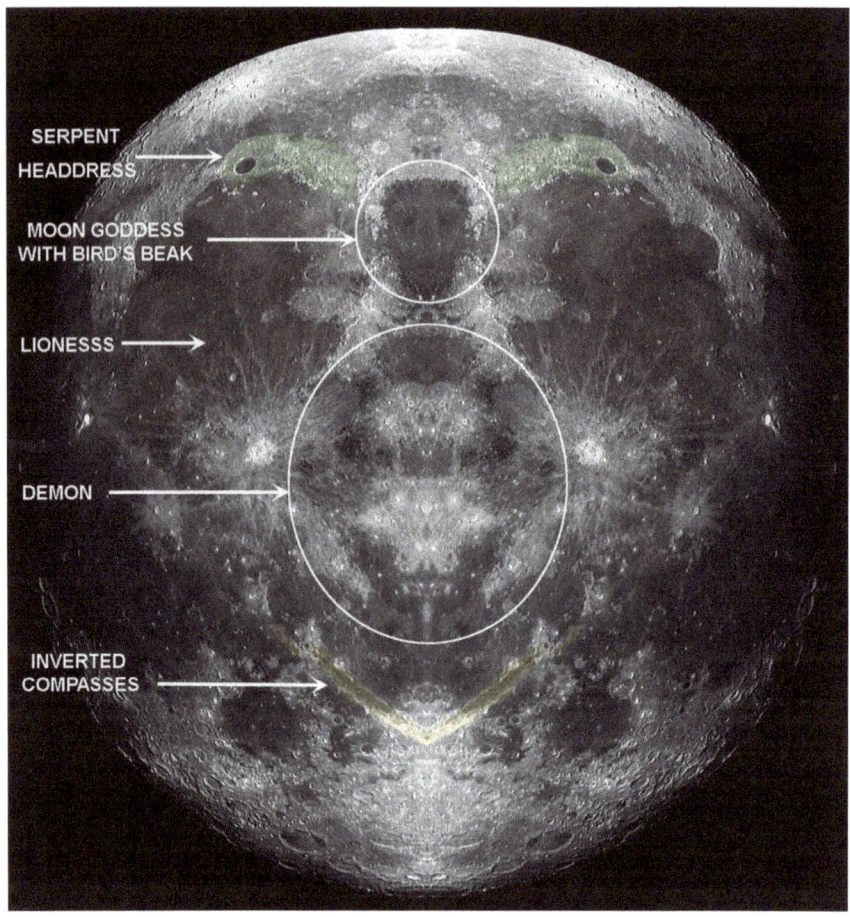

4.15 Dark Moon Goddess; Notated and false color.
Left side of demarcation line mirrored.

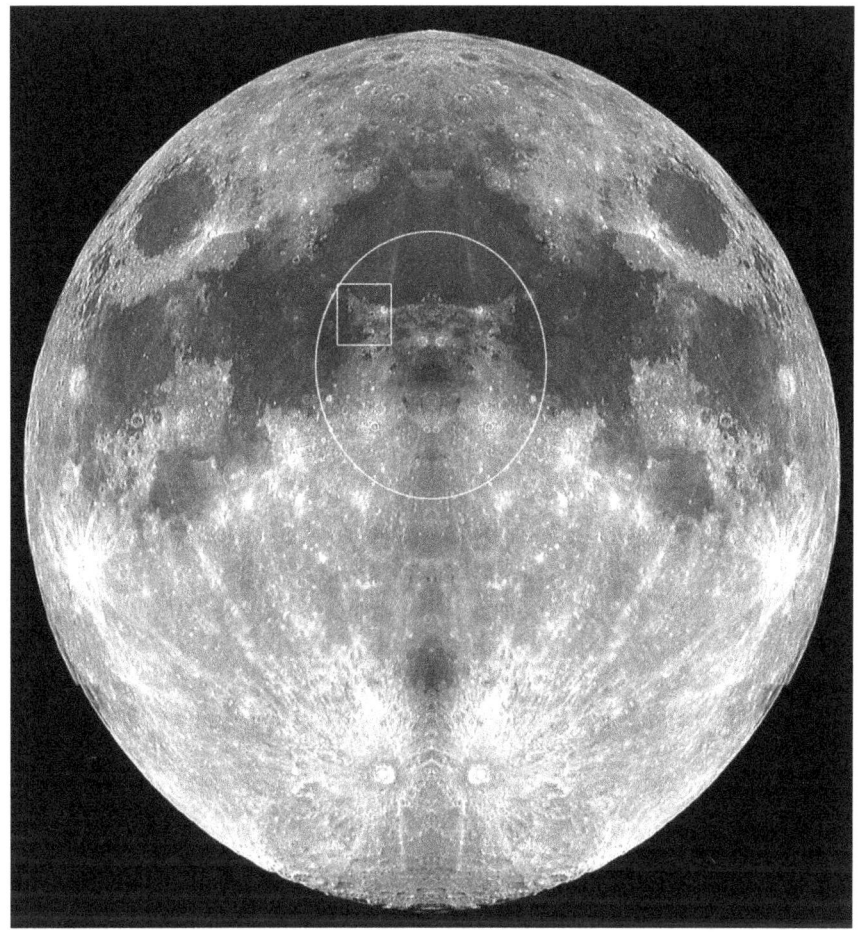

4.16 Xolotl. Right side of the line of demarcation mirrored.

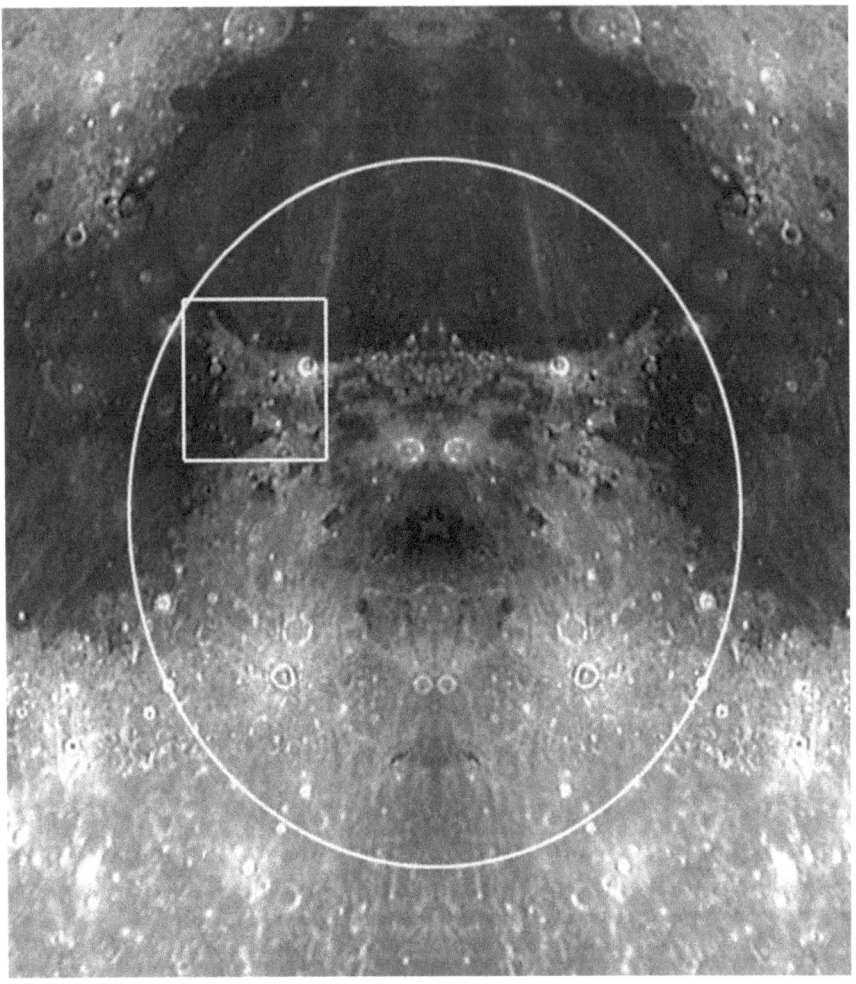

4.17 Xolotl. Detail from figure 6.16, with tattered ear highlighted.

Xolotl was also the protector of the Sun as it travelled the Underworld at night. He was associated with death and accompanied Quetzalcoatl to the Underworld to collect bones of the dead to create humans in the Fifth Sun or Fifth Creation.

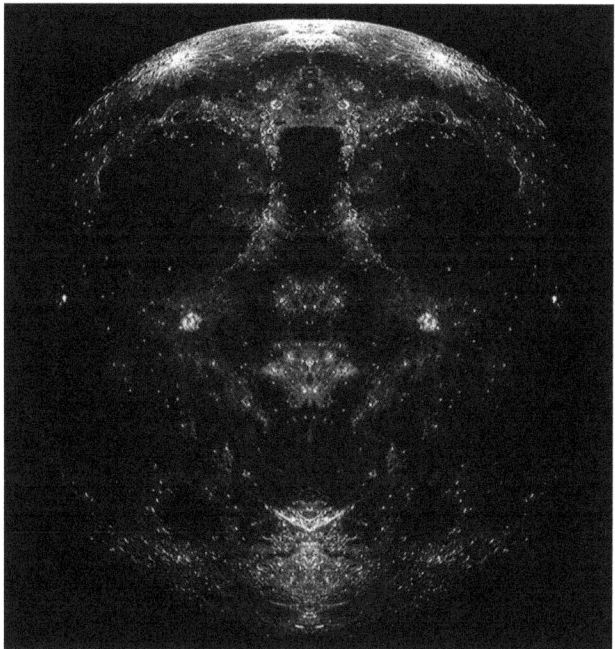

4.18 Dark Moon Goddess under high contrast. Notice her headdress shines like it is jewelled and the Moon has transformed into a skull.

In my work with the Mars images, I noticed upon occasion, I would find more 'hidden' images, or otherwise unseen portions of images, by adjusting the brightness and contrast. I tried that with the Moon goddess image. When I exaggerated the contrast, the mirrored image of the Moon goddess transformed into a skull. (Figure 4.18). A skull is a symbol of death, and as just mentioned, Xolotl helped collect the bones of the dead in the Underworld.

OUR ARTIFICIAL MOON

According to the ancient lore of many peoples, there was a time when the Moon was not yet in the sky. Aristotle wrote about it, and the ancient indigenous people of Bogota highlands in the eastern Cordilleras of Colombia refer to a time before the Moon. The Zulus of southern Africa say the Moon is hollow and was brought here long ago by two brothers of extra-terrestrial origin, Wowane and Mpanku. (Enki and Enlil?)

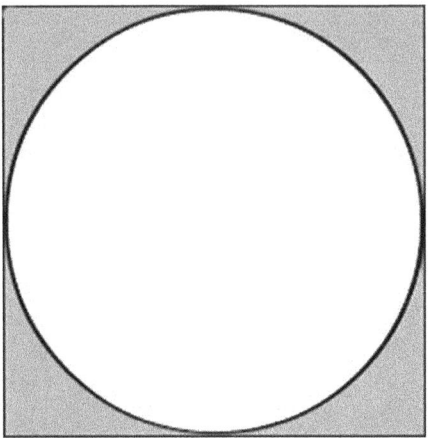

4.19 Circle in a Square; the Earth and the Moon. White is the area of the Earth, gray is the area of the Moon.

In their book, *Who Built the Moon*, Christopher Knight and Alan Butler describe the marvelous mathematical relationships built into the Moon and its position in orbit. They suggest the Moon is an artificial satellite that was put in place to allow life to flourish on Earth. In listing some of these relationships, I would note that they are not produced by physical laws. There is no scientific reason that these relationships need to exist. Yet they do.

- If the area of the circle in the square in figure 4.19 equals the Earth, then the area of the corners in grey, equals the Moon.
- 27.3 earth days is the sidereal period of the moon (the moon completes one full revolution).
- 27.3 days to complete one rotation. Since the Moon's revolution and rotation are the same only one side of the Moon is visible to us.
- The ratio of the Moon's diameter to Earth's diameter is 0.273. (The moon is 27.3 % the size of the Earth.
- 273 days = average length of a woman's pregnancy (10 sidereal months).

- 27.3 days = human menstrual cycle (This is an average as they vary from 21 – 45 days)
- The Earth and Moon's orbital periods are reciprocals. 1/27.32 = 0.0366 (366 days in a sidereal year) (1/366 =.002732) 27.32 days in one 'moonth'.
- 273 days is the length of time from the summer solstice of one year to the vernal equinox of the next year.
- Sunspots revolve about the Sun's surface in an average of 27.3 days.
- 2,730,000 is the circumference of the Sun in miles.
- The Moon is 400 times smaller than the Sun, and the Sun is 400 times further from the Earth than the Moon is, so the disk of the Moon appears to be almost exactly the size of the disk of the Sun, often giving us a perfect solar eclipse. Our Moon is the only moon in the solar system in which this occurs. It is not a coincidence.

FOOTNOTES:

1. https://www.transcendenceworks.com/moon-goddess/
2. Ibid.
3. https://www.thoughtco.com/maya-gods-and-goddesses-117947

CHAPTER V

THE EYE OF RA

"The sacred symbols of the cosmic elements were hid away hard by the secrets of Osiris. Hermes, ere he returned to Heaven, invoked a spell on them and spake these words...'O holy books, who have been made by my immortal hands, by incorruption's magic spells.....free from decay throughout eternity remain and incorrupt from time! Become unseeable, unfindable, for every one whose foot shall tread the plains of this land, until old Heaven doth bring forth meet instruments for you, whom the Creator shall call souls. Thus spake he; and, laying spells on them by means of his own works, he shut them safe away in their zones. And long enough the time has been since they were hid away."
"Men will seek out... the inner nature of the holy spaces which no foot may tread, and will chase after them into the height, desiring to observe the nature of the motion of the Heavens."

Egyptian Treatise – The Virgin of the World

1950'S GIZA PLATEAU, EGYPT

Someone once said to me that if these images on Mars are real and related to Earth's mythologies, then why don't we find them on Earth as well? I had thought of this myself at one time. We have many geoglyphs on Earth. They can be found on all the continents, except, perhaps Antarctica. The ones we are familiar with, such as those of Nasca, Peru, are not hidden like the ones on Mars. They are generally full pictographs of humans and animals in plain sight. In plain sight, that is, if you are flying over them.

In 1995, independent researcher Peter Larsen went looking for the Eye of Ra on the Giza Plateau and found it! Reasoning that if there were indeed faces and pyramids on Mars then, perhaps there were faces on Earth where there were pyramids. He found the Eye in an aerial photograph of the Giza

Plateau taken by the Egyptian air force in the 1950's. The photograph was in a book he was reading called *"The Message of The Sphinx"* by Graham Hancock and Robert Bauval.

According to Wikipedia:

> "the Eye of Ra is a being in ancient Egyptian mythology that functions as a feminine counterpart to the Sun god Ra and a violent force that subdues his enemies. The eye is an extension of Ra's power, equated with the disk of the Sun, but it also behaves as an independent entity, which can be personified by a wide variety of Egyptian goddesses... the eye goddess acts as mother, sibling, consort, and daughter of the Sun god. She is his partner in the creative cycle in which he begets the renewed form of himself that is born at dawn. The eye's violent aspect defends Ra against the agents of disorder that threaten his rule. This dangerous aspect of the eye goddess is often represented by a lioness or by the uraeus, or cobra, a symbol of protection and royal authority." [1]

5.1 Giza Plateau; Photo by the Egyptian Air Force in the 1950's.

While searching for some unrelated information online, I came across a website where Peter Larson had pointed out the Eye of Ra in the Giza photograph. However, I saw much more than just the eye; I saw the half face of a lioness. After spending years working on the Mars research, it seems I was getting rather adept at seeing half faces. I was pretty sure if I applied the mirroring process used to uncover the Mars images, I could find more at Giza. So looking for possible marker mounds for a line of delineation, as Haas and I did with the Mars photos, I created a mirrored image of the half face. This process presents one with the full face of a lioness centered amongst the pyramids (Figure 5.2).

5.2 Half face of Lioness highlighted.

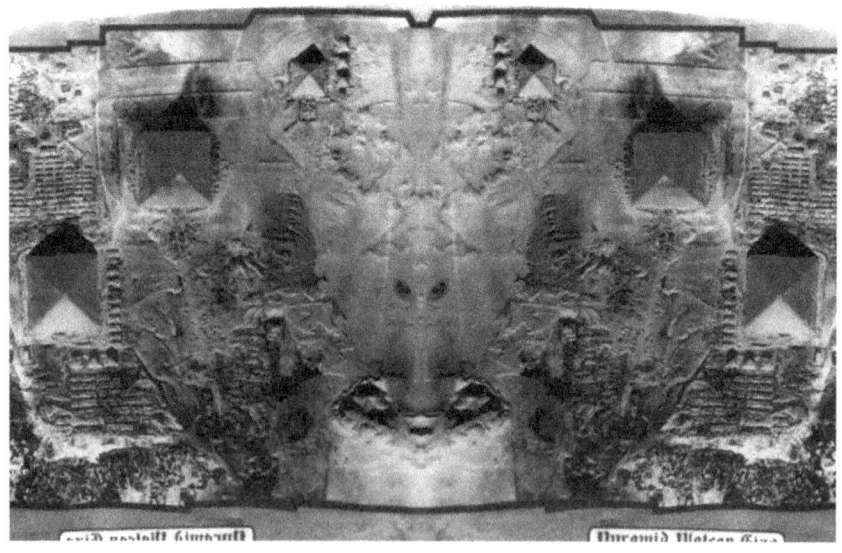

5.3 Mirrored half image of lioness.

5.4 Eye of Ra Lioness centered among the
three Egyptian pyramids. False Color.

THE QUEEN AND MORE COMPASSES

Examining the 1950's image from different angles I noticed another possible half image. I rotated the photo 140 degrees clockwise and used the peak of the small pyramid, along with two mounds on the plateau, as the demarcation line (Figure 5.5). What was created was the depiction of a deity with numerous compasses and a pair of lionesses on either side gazing at the Queen's Pyramids. Two cobras can also be seen at the deity's neckline (Figure 5.6 &5.7).

5.5 1950's Giza Plateau image rotated 140 deg.
White lines indicate location of half image.

So once again the EYE of Ra:

> *"This dangerous aspect of the eye goddess is often represented
> by a lioness or by the uraeus, or cobra, a symbol of protection
> and royal authority."*

5.6 Mirrored image of Deity with compasses
and framed by a pair of lionesses.

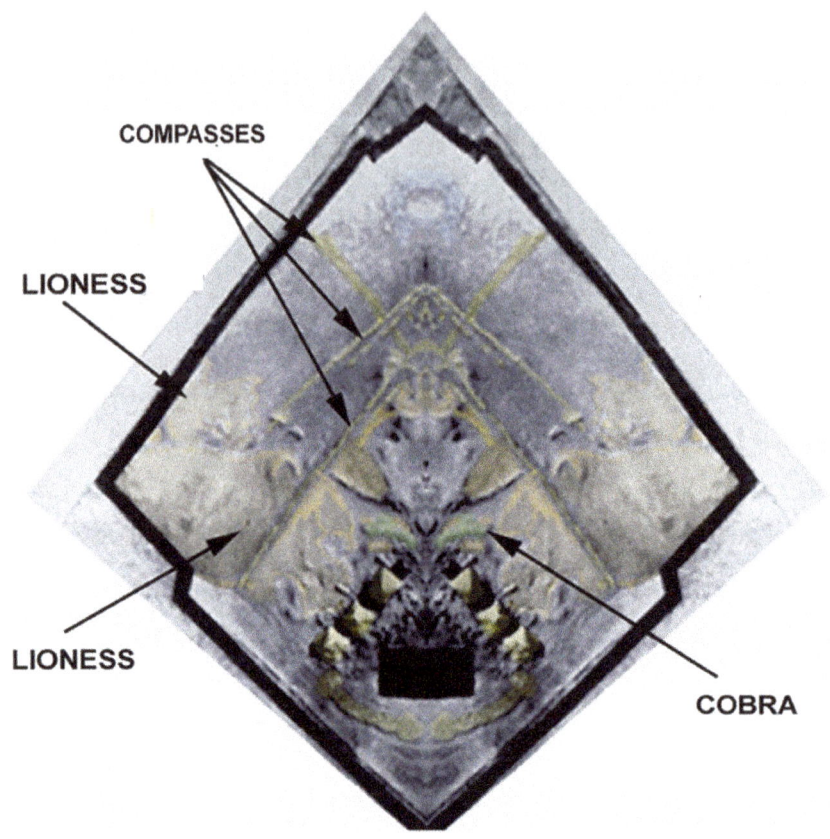

5.7 Mirrored Image of Deity on the Giza
Plateau. False color added by author.
Lioness gazes upon the Pyramids of the Queens.

Another half image was found in alignment with the peak of the great pyramid and beside the eastern cemetery.

HORUS AND THE LIONESS

Horus was the son of the Egyptian god Osiris and his wife, Isis. He, along with his parents, were three of the most worshipped gods in ancient Egypt. He was known by slightly different names and epithets, but is most recognized as "Horus the Sacred Falcon". The letter 'A' in Egyptian hieroglyphs is represented by the pictograph of a bird. In many instances it is represented by a vulture, however a vulture is a member of the falconform or falconformes order of birds.[2]

5.8 Location of half image of feline and Winged Disc emblem.

The half image is delineated through the peak of the great pyramid and along the feline half face. When mirrored the full face of a feline is presented. Due to the resolution of the photograph, it is not possible to determine the feature which forms the eye. It does, however, play a dual role as both the eyes of the feline and the wings of Horus. Twin serpents appear out of the letter 'A' in the center (Figure 5.10 & 5.11).

5.9 Mirrored feline with winged emblem.

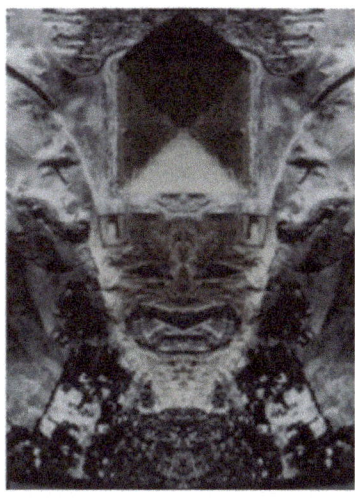
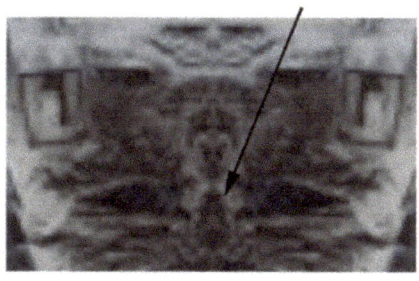

THE LETTER 'A'
AND THE WINGED DISC

5.10 Mirrored feline and winged emblem. Color added by author.
Notice the twin serpents on either side of the 'A' just
as there are in the emblem beside the disc.

5.11 Egyptian Bird Glyph Representing the Letter A.

This 1950's photograph contains more hidden images. It is unfortunate, modern disturbance and development may have destroyed historical artifacts. Perhaps it is not too late to salvage some.

FOOTNOTES:

1. https://en.wikipedia.org/wiki/Eye_of_Ra
2. https://www.britannica.com/topic-browse/Animals/Birds/ Eagle-and-Hawk-Order

CHAPTER VI

THE FEATHERED SERPENTS

Dawn blurs the boundaries of heaven and earth
Kukulkan ablaze, from spring's solar rebirth
As K'in climbs skyward, the ancient serpent fades
Where Chichen Itza's past forever replays – John Trent

THE COSMIC SERPENT

The reverence of the serpent in ancient cultures is widespread, including ancient Sumer, Egypt, Aboriginal Australia, Indigenous North American, Mesoamerica, and South and East Asia. The serpent has been used to symbolize wisdom, death, resurrection, fertility and procreation. In some cultures, spirit gods were said to sometimes appear in the form of a serpent.

6.1 Serpent images from around the world.

According to Joseph Campbell in *"The Power of Myth"*, the serpent was revered in all ancient cultures until the advent of the Hebrews. The Hebrews were male god oriented, and the subjugation of the people of Canaan included a rejection of their worship of the Mother goddess and of her symbolic association with the serpent. This marked the beginning of the inversion of the symbol of the serpent.[1]

Also revered throughout ancient civilizations was the symbol of entwined serpents. The emblem of twin or entwined serpents was a symbol of Enki, the Sumerian Creator god. Later his youngest son Ningishzidda adopted the symbol. In ancient Egypt, it was the symbol of the god Ptah, who author and historian Zecharia Sitchin claims was the same deity as the Sumerian god Enki. In Mesoamerica, Kukulkan, the central god of the Maya, was represented by the symbol of twin serpents, as was the Aztec god Quetzalcoatl. In Asia, Fuxi and Chang Jing, said to be the founders of Chinese civilization, were depicted as entwined serpents that controlled the world within the seas (as did Enki) and created the Book of Change (I Ching).[2]

6.2 Left: Entwined serpents, the Egyptian gods Isis and her husband Osiris. (A Graeco-Roman era depiction). Drawing by George J. Haas. (Image source: *The Sirius Mystery*, Temple, plate #31).
Right: Entwined serpents, Fuxi and Cang Jing (founders of Chinese civilization). The image is from a bas relief in the Han Dynasty, Wu Lianb tomb 2nd century AD. Note the mason's square in the right hand of Fuxi.
Drawing by George J. Haas. (Image source: *The Sirius Mystery*, Temple, page 294).

The caduceus, a symbol found in various forms and cultures, consists of two serpents entwined around a central axis. Wings, one assumes are representative of a bird, are often an addition to the image. This image was eventually adopted by the Greeks as the Caduceus of Hermes, and Romans as the Caduceus of Mercury, and suggests a guiding rod for souls in a quest for rebirth and eternal life.[3] The caduceus has been used to represent healing and is still in use today in various forms to signify the field of medicine.

6.3 Caduceus / Staff of Hermes

Can one conjecture that the entwined serpents, given their association with healing, were used to symbolize the double helix of DNA? Various researchers have made this connection. In his fascinating book *The Cosmic Serpent; DNA and The Origins of Knowledge*, author Jeremy Narby describes the connection between the twin serpents of ancient mythology and the double ribboned molecule of life, DNA. One of the most telling images Narby refers to comes from the Cosmic Serpent of ancient Egyptian's "Provider of Attributes" (Figure 6.4).

6.4 "The Cosmic Serpent of the
ancient Egyptians, provider of attributes."
From R.T. R. Clark, 1959, p 52.

The signs above the double serpent mean, «one» (), "several" (), «spirit, double, vital force» (), «place» (), "wick of twisted flax" (), and "water" (^^^^), Under the chin of the second serpent, is the Ankh, the Egyptian cross meaning "key of life".

As Narby states:

> "*The connections with DNA are obvious and work on all levels; DNA is indeed shaped like a long, single and double serpent, or a wick of twisted flax; it is a double vital force that develops from one to several; its place is water*". [4]

He quotes molecular biologist, Christopher Wills, who affirms:

> "*The two chains of DNA resemble two snakes coiled around each other in some elaborate courtship ritual*".[5]

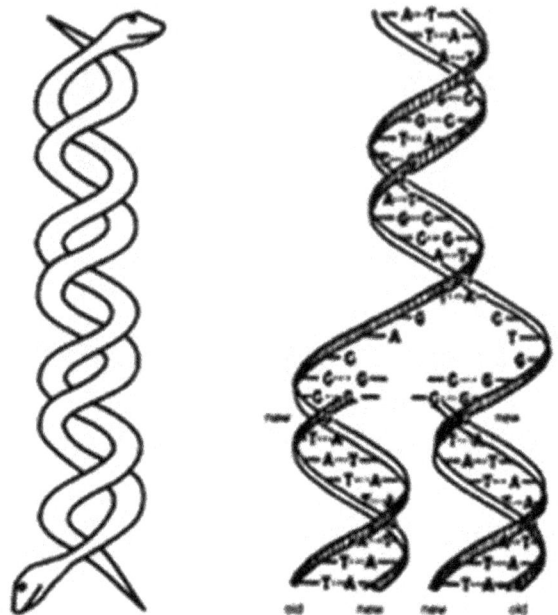

6.5 Left: DNA double helix represented as a pair of snakes.
Right: The Double Helix, James Watson, 1968.
*The Wisdom of The Genes, New Pathways in
Evolution*, Christopher Willis, 1991, 37.

The wings often associated with the caduceus present the symbolism of a bird – serpent pairing. Like the entwined serpent representations we see in figure 6.2, it is found in more than one culture. A piece called the "Birdman" is an ancient representation of a bird and entwined serpent fusion believed to originate from Southeast Iran, a product of the 5000 year old Jiroft culture.

6.6 "Birdman" 5000 + year old Bird-Twin Serpent pairing
Academia.edu

In Ancient Egypt, Horus, the son of the god Osiris and goddess Isis, was often represented as the winged disc with twin serpents. In this pairing, we once again see the dual embodiment of a bird and twin serpents.

6.7 Symbol for Horus the Sacred Falcon - Winged Disc.

KUKULKAN, THE FEATHERED SERPENT

Chapter II provided a representation on the planet Mars of the Aztec god Quetzalcoatl, also known as Kukulkan to the Maya. The half pictograph, when mirrored, presented him with a helmet, plumed with twin serpents, and a bird or flying craft above the serpents. Another interesting fact to note is that the Maya glyph for the planet Mars, called the Zip Monster by archaeologists, contains a bird - serpent pairing composed of half glyphs!

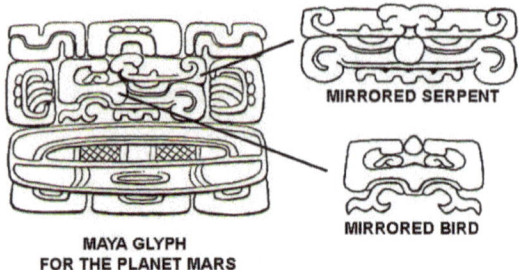

MIRRORED SERPENT

MIRRORED BIRD

**MAYA GLYPH
FOR THE PLANET MARS**

6.8 Maya Glyph for the Planet Mars consisting of half of a bird glyph and half of a serpent glyph. Drawing by George J. Haas. (Image source: *Introduction to the Study of the Maya Hieroglyphs*, by Morley, 1975, Plate 20, page 221).

In 2015, I made a stunning discovery regarding the Olmec and Maya god Kukulkan. I had been to Costa Rica a number of times since 1989, visiting with an old university friend who took up residence there in 1985. I decided it was time to pay another visit. I opened Google Earth to check on some new areas of this marvelous country to explore. As I centered Costa Rica on the screen, a half-face suddenly appeared in front of my eyes. WOW! Here was a gigantic monument of a half-face, just the type of thing Haas and I had been finding on Mars for the last two decades (Figure 6.9).

6.9 Highlighted image of half face with outstretched arm. Image source: Google Earth

I zoomed in and out a bit and captured a screen shot. Then I mirrored the half image. As I studied the completed image in front of me I was stunned. I saw shoulders, then arms, then hands…hands holding a serpent! The image was perfectly framed by the coastline. There also appears to be a bird flying upward. Undoubtedly, this was Kukulkan (Quetzalcoatl), the Maya god! The Feathered Serpent!

As I studied the image a little more carefully, I noticed that Kukulkan wears the three point crown, the symbol of royalty we pointed out with 'The Face' in chapter I. The tri-leaf crown is formed by the mirrored Gulf of Nicoya and another small portion of the continental shelf. I wondered for a moment if this image was oriented north-south, since alignment to the poles, the rising Sun at the solstices or the equinoxes, was an important aspect of ancient monuments. It didn't take me long to realize this was not a north-south alignment, but it did suddenly strike me that the pyramid of Kukulkan in Chichen Itza, Mexico was somewhere to the north. Sure enough, as I extended the demarcation line of the half-face and continued it northward for over 1300 kilometers, it intersected the Pyramid of Kukulkan!

The Pyramid of Kukulkan is actually three temples or pyramids built atop one another. All are believed to have been built during the first millennium C.E. The edifice is stepped with nine layers or platforms. Each side has a central staircase with 91 steps and with four sides that totals 364 steps while the top platform represents the 365th day of the year. Of course, the pyramid is most famous for the equinox serpents that descend the stairway twice a year on the day of the spring and fall equinox. During these days, the sun projects seven triangles of light that slowly slide from top to bottom and the last ray of light shines onto the head of the feathered serpent at the base of the stairway. The phenomenon lasts approximately 3 hours before sunset. It is said that Kukulkan appears twice a year to check on the well-being of the people and the crops before descending into the Underworld through the Cenote that lies beneath the pyramid.

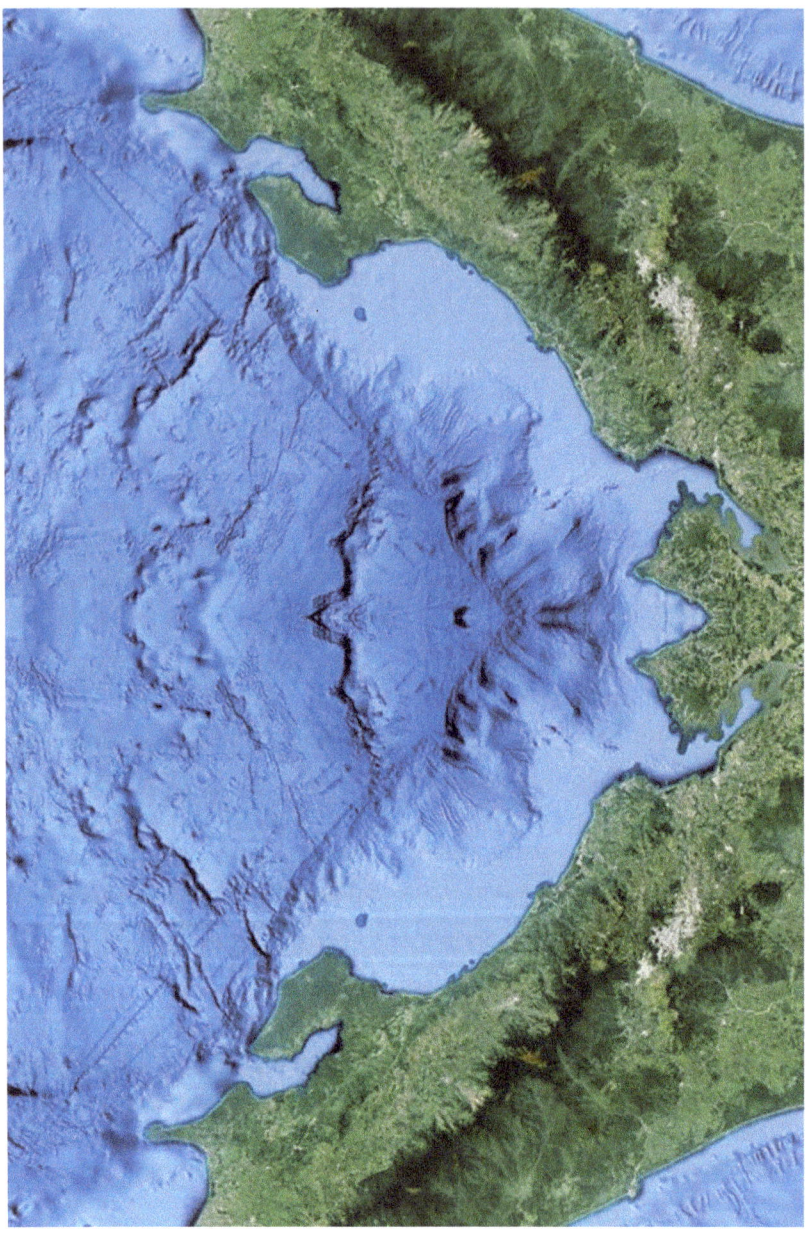

6.10 Mirrored image of Kukulkan with bird and twin serpents.

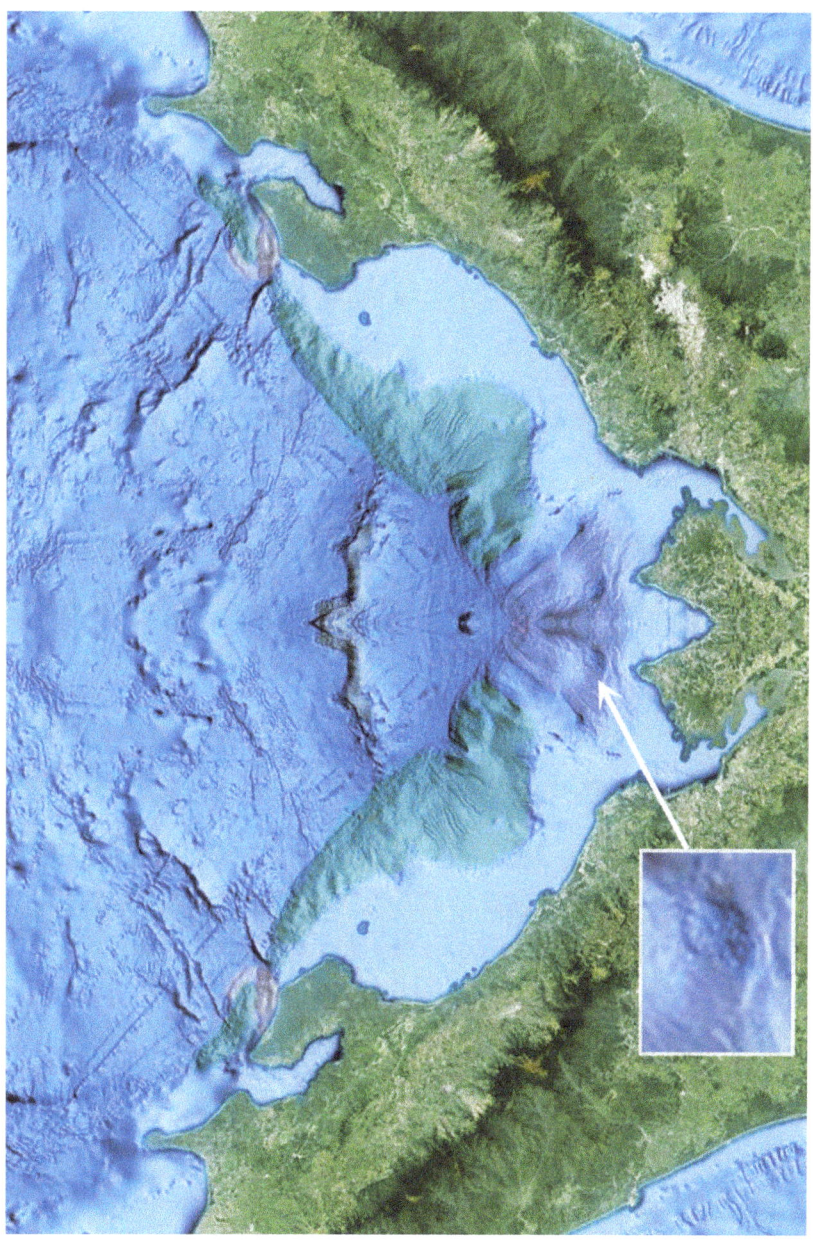

6.11 Mirrored Kukulkan (Quetzalcoatl) with bird and twin serpents. False color added by author.

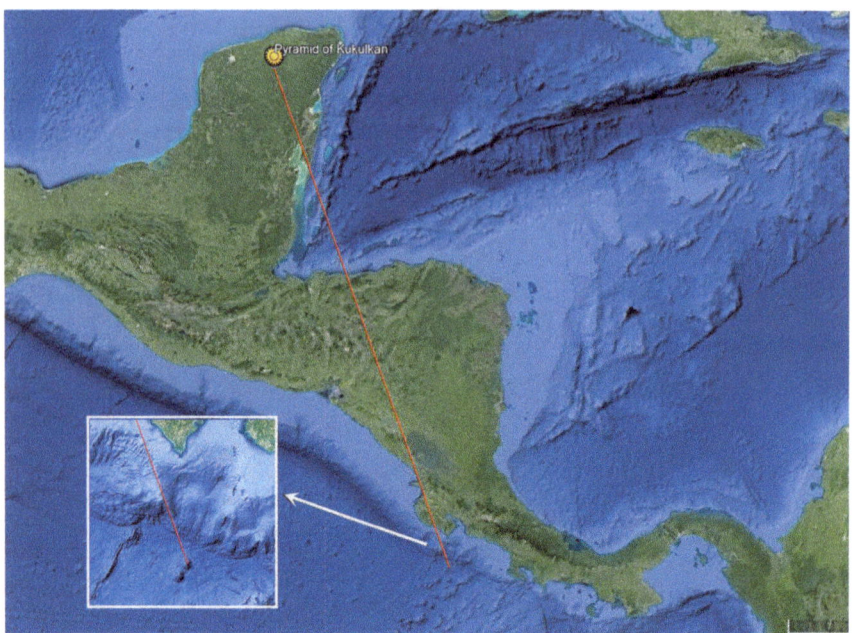

6.12 The demarcation line used for mirroring the half image is determined from two mounds on the bridge of the nose and a third submarine mound on the ocean floor. When the line is continued northward for over 1300 kilometers, it eventually intersects the Pyramid of Kukulkan in Chichen Itza, Mexico.

I knew this changed everything. This was exactly the kind of evidence that vindicated our Mars research. Not that we should have needed any, the Mars work stands on its own. But here … here we have a monument on our own planet that is roughly four kilometers from the bottom of the ocean to the top of the continental shelf and stretches for 160 kilometers to the south along the edge of the shelf.

Notice that the Kukulkan image appears to be blowing something, possibly a seed. In Maya mythology, Kukulkan created all things with his breath. In Hebrew mythology, the Bible states: "By the word of the lord were the heavens made; and all the host of them by the breath of his mouth".[6]

6.13 Pyramid of Kukulkan in Chichen Itza,
Mexico with the Equinox serpents.
Serpent body created from sunlight and shadows.

BIRDMAN (ROCKET MAN)

After stumbling across the Kukulkan (Quetzalcoatl) image, I began
scanning the area for more. I certainly did not expect the findings to be
even larger in size, but they were.

The next image is massive (Figure 6.14). It is a half image I nicknamed
"Birdman" because he has outstretched wings. However, he could be called
"Rocket Man" since there appears to be a portrayal of the expulsion of
fire and exhaust below him. A continuation of the arced extension of the
serpent in the Kukulkan image forms a helmet and obviously expresses the
serpent aspect of the Feathered Serpent. The image is oriented perfectly
north-south and extends approximately 900 miles or 1400 kilometers in
length and roughly 500 miles or 800 kilometers from the wing tip to the
center of the face. The demarcation line at the center of the face is at 83.45
degrees W longitude, but I have yet to find a significance to this number.
However, following the line northward, just as we did with the Kukulkan
line of demarcation, it runs roughly one degree longitude away from
a possible archeological discovery. In 2001, Canadian marine engineer
Pauline Zalitzki and her husband Paul Weinzweig discovered underwater

evidence of pyramids just off the western tip of Cuba at 84.83 degrees west longitude and at a depth of 600 – 750 meters (2000 – 2460 ft.).

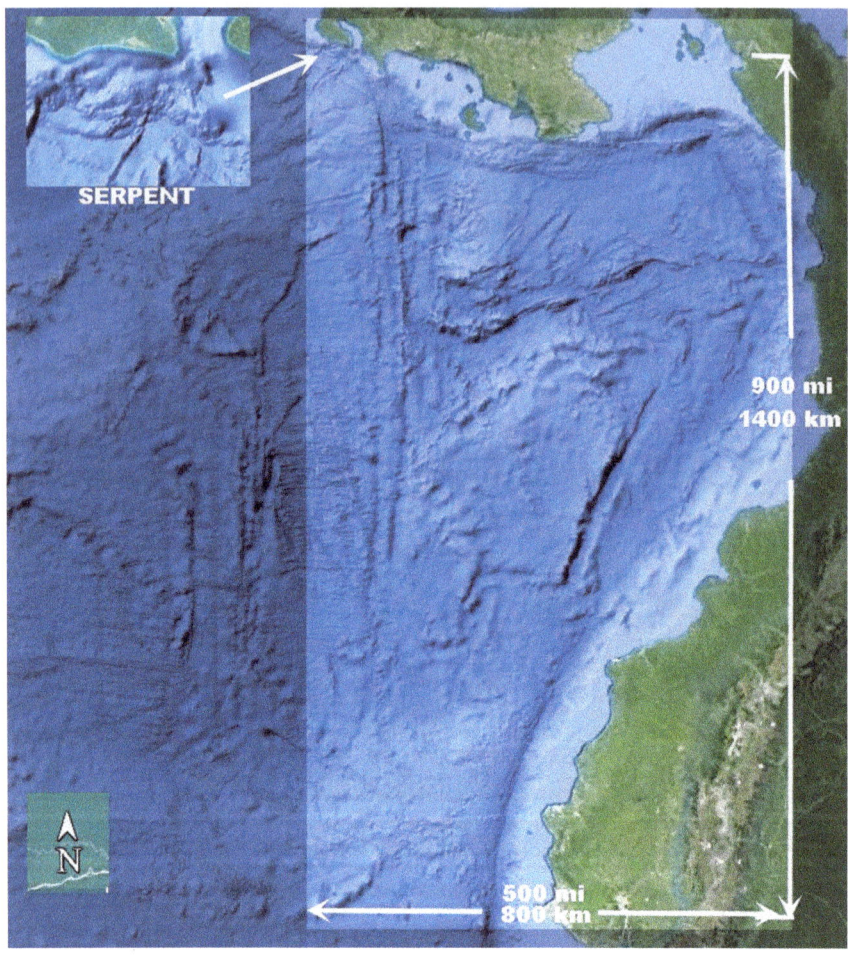

6.14 Context location of the half image of 'Birdman'. Note the location of the serpent from the Kukulkan image from Google Earth.

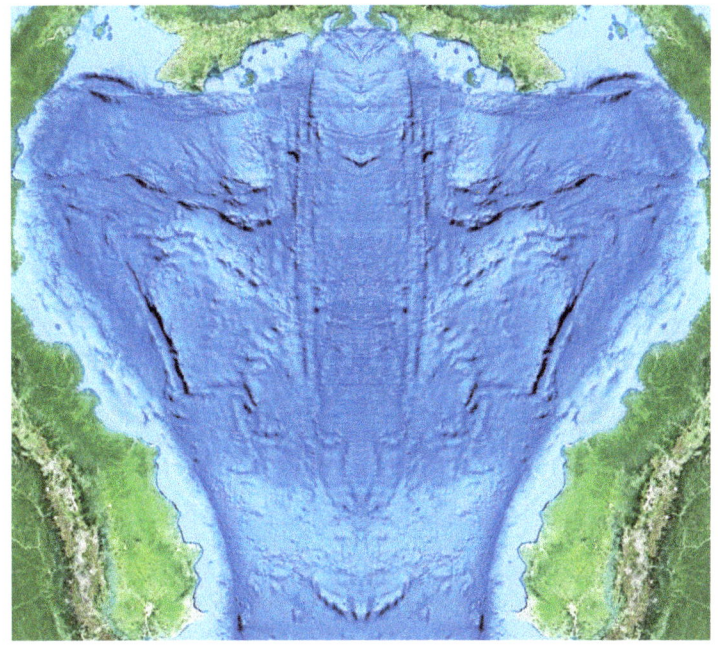

6.15 'Birdman'. Mirrored half image from figure 4.14.

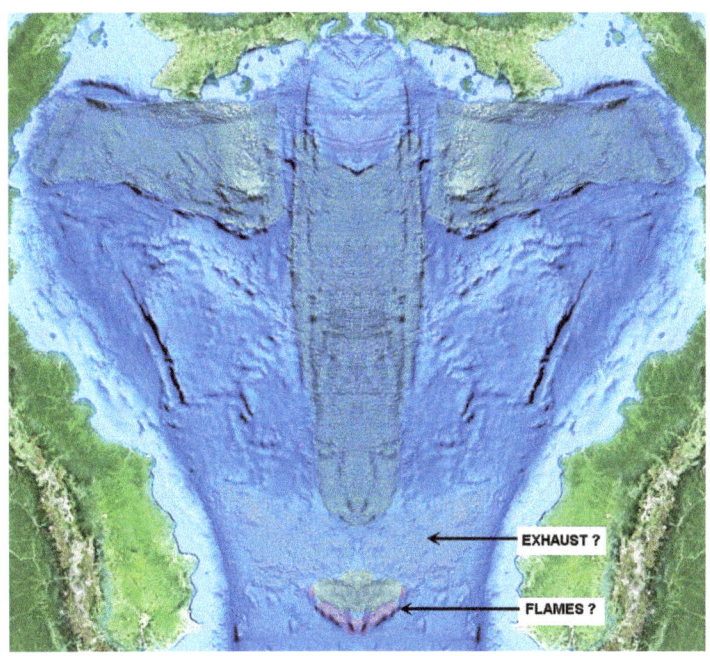

6.16 'Birdman'. False color added by author.

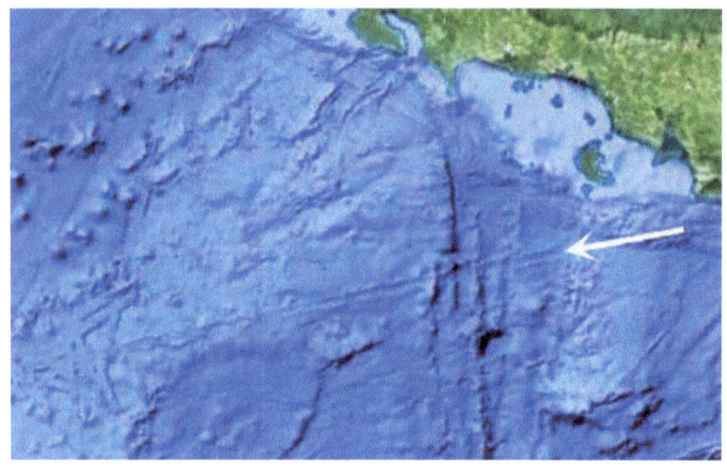

6.17 New line on ocean floor.

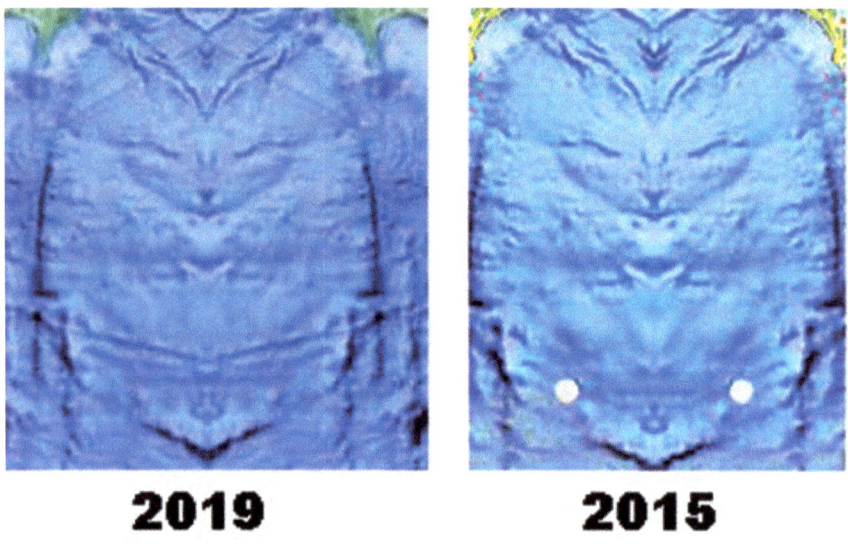

2019 **2015**

6.18 Birdman face from two different years. The isolation images of the face show that there has been activity in this area since 2015. Notice the line over the mounds which form the nostrils.

I noticed that a change in the image had occurred sometime in the last four years. There is now a very straight, etched line passing through the nose area of the image. If one looks at the ocean floor in Google Earth, one sees an abundance of perfectly straight lines, especially off the west coast of the Americas. We know from the images shown here, some of these lines are artwork. However, we must ask the question: is someone mining the ocean floor?

THE GREAT WINGED DISK

As previously mentioned, the three most worshipped gods of ancient Egypt were Osiris, his wife Isis, and their son Horus. Osiris was associated with the constellation Orion, while Isis was identified with the star Sirius. Their son, Horus, the Sacred Falcon, was associated with the stars, the Sun and the Moon. He is also considered the god of death and resurrection and of the sunset and sunrise.

The ancient Egyptian holy city of Edfu is dedicated to Horus. He is said to have established a foundry of "divine iron" where he kept "the great winged disk that could roam the skies". According to author and historian Zecharia Sitchin, the Egyptian text declared:

"When the doors of the foundry open, the Disk riseth up." [8]

I now take you from the massive image of "Birdman" to an image that could be termed colossal. It is also at the bottom of the Pacific Ocean, and like the Birdman, it is oriented perfectly north-south. The demarcation line for the hidden image of "The Great Winged Disk" is 109.5 degrees west longitude and stretches from roughly 3 degrees south latitude to about 60 degrees south latitude. The half image is roughly 4,500 kilometers or 2,800 miles in width and 6000 kilometers or 3,700 miles in length. When mirrored, the image formed is what we have come to call a flying saucer. The image of the disk also shows a person within the cockpit and another one in the foreground. At the top of the image, North America becomes a smiling face, complete with eyes, ears, nose, and mouth. At first, I thought this being in the foreground was hovering, but on closer inspection, he appears to be wearing swimming fins on his feet. He also seems to have water (tears?) flowing from his eyes.

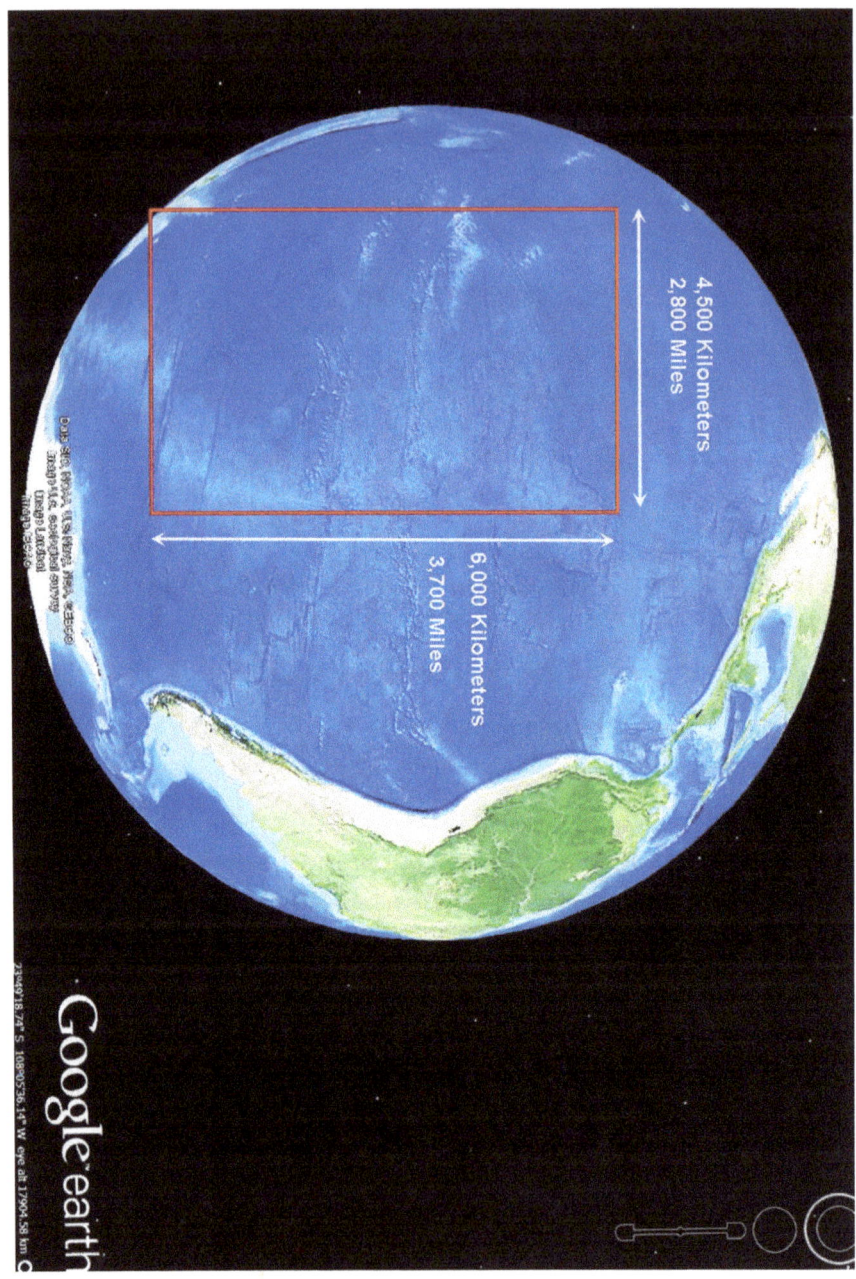

6.19 Location of the half image of 'The Great Winged Disk'. Image source, Google Earth.

6.20 Mirrored image of 'The Great Winged Disk'.
Also, notice the smiling face above.
(Rotate image 90 ° counter clockwise for viewing).

6.21 Top: Mirrored Half Image of The Great Winged
Disk, Bottom: False Color added by author.
(Rotate image 90 ° counter clockwise for viewing).

The numerical values of latitude and longitude were designated in 1851, but I still hoped to see some significance for the saucer's demarcation line. Consequently, I was initially disappointed in the 109.5 number for the north-south demarcation line. The number seemed to be merely a random figure. However, after some further research, my thoughts would change.

In 1983, The Independent Mars Investigation, a group of researchers spearheaded by science journalist Richard C. Hoagland, began using the 1976 Viking Orbiter images to study the structures in Cydonia, Mars. In the design and alignments of the structures, they found mathematical concepts pertaining to tetrahedral geometry.

A tetrahedron is the simplest form of the platonic solids, containing four equal sides and four equal angles. If a theoretical tetrahedron were placed inside the Earth with one point at the South Pole, the other 3 points would touch the Earth's sphere at 19.5 degrees north (visa versa with the North Pole). The researchers noted at roughly that latitude, 19.5 degrees north or south, a number of geophysical features appear throughout the solar system. There are the Hawaiian Island's shield volcanos on Earth, active volcanic complexes on Venus (Alpha and Beta Regio), the Great Red Spot on Jupiter, the Great Dark Spot on Neptune, and the largest geophysical structure on Mars, the massive shield volcano, Olympus Mons.[9]

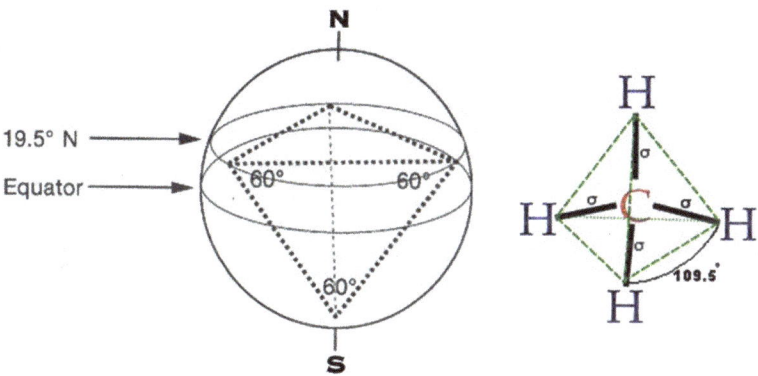

6.22 Left: Tetrahedron in a sphere; Right: Tetrahedral molecule of one of the simplest organic compounds, methane (one central carbon atom and 4 hydrogen atoms).

A little investigation into the demarcation line of 109.5 for the saucer, revealed that 109.5 is the bond angle for organic molecules that are tetrahedral in shape (figure 4.22).

*In **tetrahedral molecular geometry**, a central atom is located at the center with four substituents that are located at the corners of a tetrahedron. The bond angles are cos–1(–⅓) = 109.4712206...° ≈ 109.5° when all four substituents are the same...*[10]

So, it seems that once again, our attention is being drawn to tetrahedral geometry, but why? The Independent Mars Investigation team believe their findings point to a domain of hyper-dimensional physics. The physics and mathematics involved in their study is beyond the scope of this book, but we will be pointed back to geometry in next chapter.

FOOTNOTES:

1. Joseph Campbell with Bill Moyers, *The Power of Myth*, (New York, First Anchor Books Edition, 1991) 55
2. Robert Temple, *The Sirius Myst*ery (Rochester, Vt. Destiny Books, 1998), 294.
3. Joseph Campbell, *The Mythic Image* (New York: MJF,1974) 283
4. (Narby, Jeremy, *The Cosmic Serpent: DNA and the Origins of Knowledge* (New York: Tarcher/Putnam, 1999). 102
5. Ibid: 92
6. Hebrew *Bible*, Psalms 33:6
7. https://en.wikipedia.org/wiki/Cuban_underwater_city
8. Zecharia Sitchin, *The Cosmic Code*, Book VI of The Earth Chronicles, (New York, Avon, 1998), 172
9. Richard C. Hoagland, *The Monuments of Mars* (Berkley, Calf. North Altlantic Books, 1992), 353
10. *Brittin, W. E. (1945). "Valence Angle of the Tetrahedral Carbon Atom". J. Chem. Educ. **22** (3): 145. Bibcode:1945JChEd..22..145B. doi:10.1021/ed022p145.*

CHAPTER VII

STARMAN AND THE GEOMETRY OF LIFE

"If you want to understand the Universe,
think Energy, Frequency and Vibration"
- Nichola Tesla

STARMAN

Finding the saucer and the Quetzalcoatl images on the bottom of the ocean spurred me on to search for more. In some areas of the ocean floor, one can find an enormous number of straight lines that criss-cross at different angles. I am certain some of these lines can be attributed to faults, fractures, and seafloor spreading. Others, as I found, were part of intentional designs. Zooming out on Google Earth to view most of the Pacific Ocean from a global perspective, I noticed a very large eye-shaped form. I searched for a spot that would locate this eye within a proportional half face. Finding a likely location, I then looked for any possible markers created by the artist(s) that may define a line of demarcation. Satisfied I had a possible location for this line, I created a mirror image of the half face. Putting the two halves together, I was in awe at what I saw; a planet-sized face adorned with a star-shaped headdress (Figure 7.3)..

After my initial astonishment began to diminish, I studied the perfect almond shaped eye and its circular iris. I was drawn to the jewelled forehead, just above the eyes. I wondered what geological formations were creating this design. Were they islands? No, a closer examination showed that they were submarine mountains. My eyes wandered again, this time to the almost equidistant lines that seemed to emanate from the head, giving a 'shooting star' effect to this miraculous image.

When I checked the latitude and longitude grid lines, I saw that this image was not aligned north-south, but I was intrigued that the approximate center of the image was very close to the point where the Earth's equator crosses the antemeridian. The antemeridian being the continuation of the prime meridian as it appears on the opposite side of the globe. The Prime and Antemeridian separate the eastern and western hemispheres while the equator separates the northern and southern hemispheres. It seemed to me to be too close for random chance. By moving the demarcation line so that it would line up perfectly to the intersection of the two lines created a face that was disproportionate. I was convinced my original line of demarcation was accurate and attributed the fact it was slightly off the intersection of the two grid lines was due to the movement of the Earth's crust since the image was created. Perhaps, though, there was another reason for this alignment that I had not yet discovered.

Obviously the equator and the antemeridian cross at a 90 angle, but when I measured the two angles created by the line of demarcation for the half face, I realized a perfect right-angle triangle was created with 30, 60, and 90 degree angles (Figure 7.2). I don't think this is by chance. This underlying geometry should come as no surprise as geometry has been a critical aspect of the Cydonia research since Vincent Dipietro and Gregory Molenaar began their research into the Face on Mars and the D&M Pyramid in the late 1970's.

7.1 Half face in Pacific Ocean. Context showing line of demarcation.

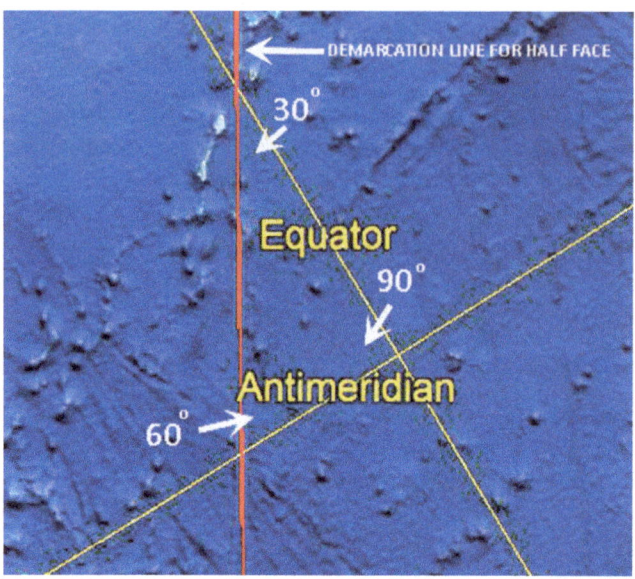

7.2 Line of demarcation for the half face along with the equator and the antemeridian lines create a perfect right-angle triangle.

If you recall the mirrored humanoid side of the 'Face on Mars' in chapter I has a nose ornament obscuring the area where the nose would be? Looking closely at where the nose should be on this face, we are again presented with a nose ornament, this one of familiar design, the square and compasses! As my eyes scanned the image once more, I realized the symbolism of the square and compasses had been echoed within the jeweled headdress as it contains a circle and a square (Figure 7.4).

I think it is appropriate here to revisit what was discussed in regards to the square and compasses symbol from chapter II.

The square and compasses are tools of geometry and architecture but are well known as the most identifiable symbol of the Freemasons, an organization that has root connections to the Megalithic Master Builders and ancient Egypt. They are also symbols of opposing elements; the circle and the square. The duality extends further, in that the circle represents the heavens, and the square represents the Earth.

As J.E. Cirlot states in *The Dictionary of symbols*:

"An emblematic representation of the act of creation, found in allegories of geometry, architecture and equity. By its shape it is related to the letter A, signifying the beginning of all things. It also symbolizes the power of measurement, of delimitation."

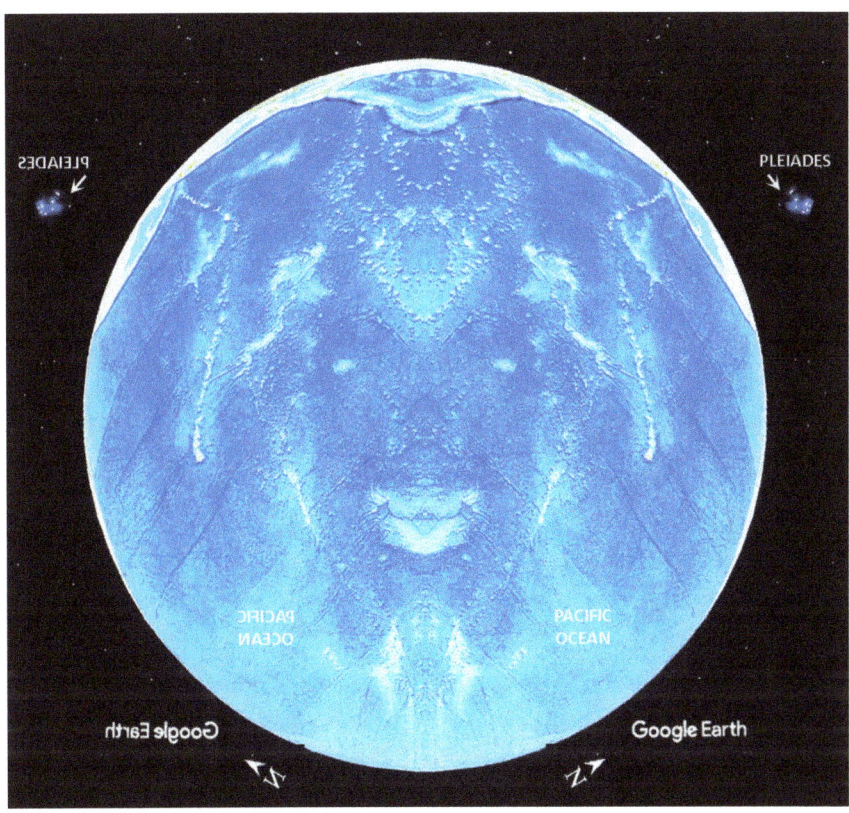

7.3 Starman. Mirrored Image of figure 7.1.
The stars in the background change with the revolution of the planet.
The timing of this just happens to capture the Pleiades cluster.

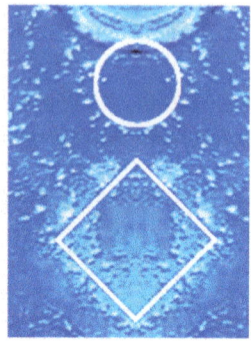

7.4 Square and Compasses as a nose ornament,
with a circle and a square in headdress.
(From figure 8.3)

When George Haas and I were assembling the material for our first book, we were trying to understand why the images or geoglyphs we were finding on Mars were presented in half and bifurcated form. We understood that the Maya had produced art pieces in this form, but why? We learned that they had a strong belief in duality and wrote about that in our books. Whether the information in this art work is a message for us or just their trademark, or emblem, as they travel the universe, it must have importance in the larger scheme of things.

THE SQUARE, THE CIRCLE AND UNIVERSAL DUALITY

There is a subject known as sacred geometry which stems from an ancient belief that a God created the world according to a geometric plan. The belief can be found in some religions such as Hinduism, Christianity and Islam. My study of geometry ended many years ago in the 12th grade so I cannot delve into the subject to any great depth. There are however, some basic aspects of which many readers may not be aware.

In chapter III, I briefly touched on the symbolism of the square and the circle in reference to the Great Pyramid in Giza, Egypt, where the height equals the radius of a circle whose circumference equals the perimeter of the pyramid's square base.

The circle and the square symbolize duality. The circle represents many things, universal unity, a whole, oneness, the masculine, the spirt, and heaven. The square represents the whole poised for manifestation. It creates the four cardinal directions and allows the comprehension of space and opposing forces; it represents the feminine, matter, and the Earth.

Given a circle having a center at point 'O', a radius of 1 unit, and a diameter A A' and B B' crossing at right angles in the center (Figure 7.5a), one can draw two circles within the larger circle using the center point radius B O (Figure 7.5b). This creates two circles whose combined circumferences equal the circumference of the large circle. However, the area of the two circles added together is only half of the large circle. One has become two. Also created is the basis for the ancient Chinese symbol of Yang Yin, seen in figure 7.5d. The Yang Yin also represents duality, and in the *Dictionary of Symbols*, Cirlot describes the Yang Yin as the "dual distribution of forces, comprising the active or masculine principle, Yang (white) and the passive or feminine principle, Yin (black). Each, however, contains within it a circle of the other to symbolize that every mode must contain within it the germ of its antithesis. [1]

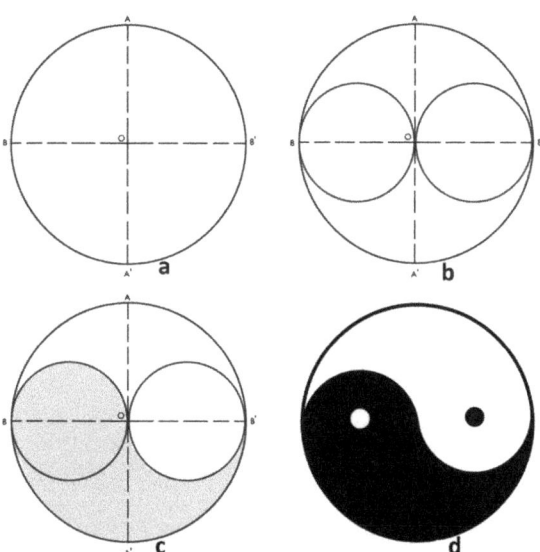

7.5 a) Circle with bisecting lines. b) Two circles inscribe within the original. c) The Yang and Yin formed from the three circles. d) The ancient Chinese symbol of Yang Yin.

Using A as the point to set the compass and drawing an arc that is tangent to the two smaller circles, and then drawing another arc, this time using point A' to set the compass, a vesical is formed (figure 7.5 c). This vesical divides the diameter A A' of the original circle into a segment A E, which happens to be Phi. Like Pi (π), Phi (Φ) is an irrational number (1.6180339…). As one shall see later in this chapter, Phi plays an important role in many life forms. If the radius of the original circle is considered to be 1 unit, Phi is found in the distance between the intersection of the lower arc of the vesical and the vertical diameter of the circle shown as line A E in figure 8.6. The inversion of Phi, 1/Φ, is found within the radius, as is 1/Φ². If one encloses the original circle in a square, where the sides of the square are tangent to the circle, and uses the center of the original circle O, one can create another circle tangent to the tip of the vesical (figure 7.7).

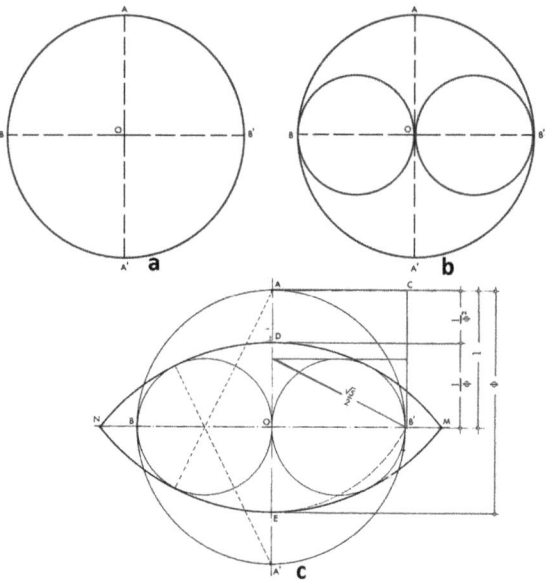

7.6 The creation of the vesical, and the demonstration of Phi within the geometry.

Since the radius of the original circle was 1 unit, the perimeter of the square is 8 units; and the circumference of the large, or outer circle is 7.993. Because Pi is an irrational number, the perimeters of the two shapes can never be exactly the same when created as such.

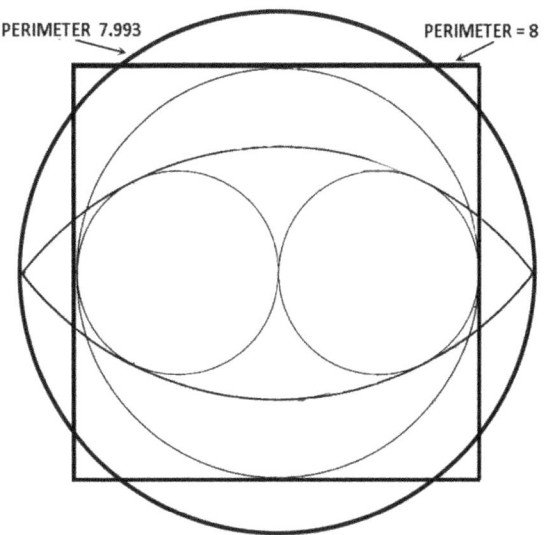

PERIMETER 7.993

PERIMETER = 8

7.7 Circle and square with almost identical circumference and perimeter created using only compasses and square.

The Maya had a similar view of duality, which is expressed by the Hunab Ku, a square within a circle (Figure 7.8). To form the Hunab Ku, the Maya would first form a square and quarter it to obtain the center. Then by placing a cord at the center, stretching the cord along a 45-degree angle to the corner of the square, and running the cord completely around the square, they were able to form a perfect circle.[2] To the Maya, this very geometric shape of a circle and a square established the concept for the duality of the universe. Their myths and bifurcated glyphs provide evidence that all aspects of their world and their conception of humanity were composed of paired, complementary deities.

Chania is a city on the northwest coast of the island of Crete. It was formerly known as Cydonia (or Kydonia), an ancient city-state said to have been founded by King Cydon, a son of Hermes or Helios (Apollo) and of Akakallis, the daughter of King Minos. An ancient coin from this city has on its reverse side, a geometric diagram of a quartered square with a diagonal marker in its lower left-hand corner. This diagonal marker signifies the act of drawing a cord from the center of the square to its lower corner and pulling it around the outside of the square, thereby forming a perfect circle. Because of the diagram's unique design, one could say it is a demonstration of the Maya Hunab Ku, with the outer rim acting as a circle. In this case, unlike the process in figures 7.6 and 7.7, the circumference and perimeters are not equal.

7.8 Left: Maya Hunab Ku, Square in a Circle;
Right: Cydonia, Crete coin with god, Apollo, and square in a circle

The obverse side of the coin features the head of Apollo, the Sun god who was known as "the destroyer" in Homer's Iliad. Apollo was known for his prophetic abilities and was the god of many things, including the arts, music, and mathematics. His appearance on the face of this coin may be another reference to the art of sacred geometry. In the *Iliad*, Homer credits both Apollo and Poseidon with the building of the walls of Troy.[3] Furthermore, the Greek poet Callimachus described Apollo as a great builder. In one of his poems he proclaims that Apollo "delights in the construction of towns of which he himself lays the foundation."[4] In this context, where Apollo is portrayed as the Great Architect of Cydonia on a commemorative coin, his attributes are set in opposition: he is seen as both "the destroyer" and "the builder."

Like the simple graphic design of the Hunab Ku, the quartered-square diagram encodes the duality of the circle and square, and further establishes a common analog to the duality of opposing forces. The importance of this symbol to the Maya is demonstrated by the venerated accouterments found with the great Maya King Pacal of Palenque; a jade cube in his right hand and a jade sphere in his left hand. (Figure 7.9.)

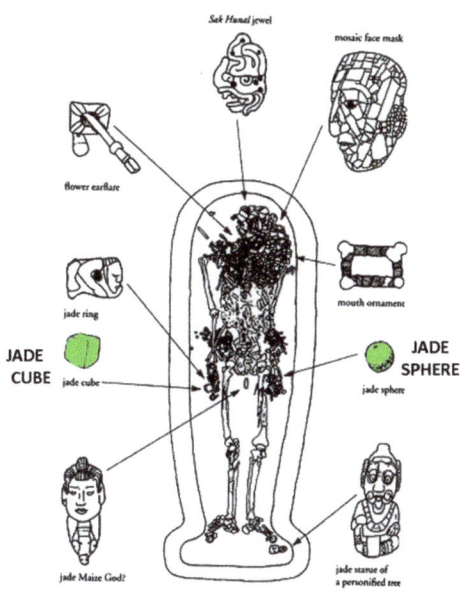

7.9 Sarcophagus of the Maya King Pacal
(his remains and its accouterments).
Note the jade cube in the right hand and
the jade sphere in the left hand.
(Green highlight add by the author). Drawing by Linda
Schele, © David Schele, Courtesy of Foundation for the
Advancement of Mesoamerican Studies, Inc. 2002.

VIBRATIONAL GEOMETRY AND PERCEPTION

In 1924 Louis Raymond de Broglie postulated that matter was both a particle and a wave. The wave-particle duality concept is known as the de Broglie hypothesis, and is a central part of the theory of quantum mechanics.[5] In simple terms, when one takes what we see as solid matter, and breaks it down to its smallest components, what is left is an extremely small packet of energy in the form of a vibrating wave. This property was experimentally demonstrated in 1927 and won De Broglie the Nobel Prize for Physics in 1929.

It has been said that if there was ever a universal language it would be expressed in math, geometry, energy patterns, and frequency. Geometry is the branch of mathematics that deals with the measurement, properties, and relationships of points, lines, angles, surfaces, and solids.[6] While doing my researching I came across some very interesting books and websites that take a different look at what we usually understand about shapes and our visual world. Two of these websites were sonicgeometry.com and vesica. org where I became fascinated with shapes and sounds.

In looking at the basic geometric shapes, one finds that the simplest form of a two dimension shape is a triangle. A triangle has 3 sides and 3 angles that total 180 degrees. A circle has no angles, but we divide it into segments of 360 degrees. A square has 4 sides and 4 angles that add up to 360 degrees. A pentagon has 5 sides and 540 degrees, a hexagon, 6 sides and 720 degrees, a septagon, 7 sides and 900 degrees; and an octagon, 8 sides and 1080 degrees. Notice that the digits of all these angles add up to 9.

$$180 \quad 1 + 8 = 9$$
$$360 \quad 3 + 6 = 9$$
$$540 \quad 5 + 4 = 9$$
$$720 \quad 7 + 2 = 9$$
$$900 \quad 9 + 0 = 9$$
$$1080 \quad 1 + 8 = 9$$

Let us combine geometry and energy, energy in the form of sound since sound is the translation of vibration into noise. Vibrations are measured in seconds and expressed in hertz, where one vibration per second is one hertz.

The normal range of human hearing is from 20 hertz to 20,000 hertz. Currently, music is based on tuning that uses a pitch of 440 hertz for note 'A' above middle "C" on a piano. This practice was officially adopted in the 20th century. All notes are created from this frequency, each half tone being equidistant from the next. This is termed equal temperament. An octave is the same note but with a frequency two times higher or one half lower (figure 7.10).

The Greek scholar Pythagoras (570 – 495 BCE) developed a tuning system based on mathematics, in which the intervals are whole numbers and based on a ratio of 3:2. This ratio, called the perfect fifth, is considered one of the most pleasant and easiest to tune by ear. Octaves still have the same 2:1 ratio in the Pythagorean system as the equal temperament system. The Pythagorean system was prominent until the mid-16th century. Between the mid-16th century and the 20th century different frequencies were used for the basis of tuning. In 1988 major opera singers of the world, in conjunction with the Schiller Institute, lobbied to have the 440 system changed to the 432 system, since the 432 frequency was considered a better tone to maintain natural registers of the voice. Some musicians have begun to change the pitch of their tuning from 440 to 432, though many orchestras throughout the world use a pitch of anywhere from 440 to 452 hertz. The 440 equal temperament system may have better consonance at higher octaves, but by using the number 432, we see the magic of the Pythagorean system (Figure 7.11).

Comparing the two charts in figure 7.10 and 7.11 one will notice that, other than notes A and B, there are no whole numbers in the chart in figure 8.10 based on A = 440. The Pythagorean chart, based on A= 432 is all whole numbers.

Did you ever wonder what a triangle sounds like? Probably not. We don't usually think of sound as a shape, but if one uses the Pythagorean system and applies a corresponding frequency to the angles of the basic geometric shapes, some very interesting things happen.

Note Vibrational Frequency of Each Octave

C	130.82	261.63	523.26	1046.52
C#	138.6	277.18	554.36	1108.72
D	146.83	293.66	587.32	1174.64
Eb	155.57	311.13	622.26	1244.52
E	164.82	329.63	659.26	1318.52
F	174.62	349.23	698.46	1396.92
F#	184.99	369.99	739.98	1479.96
G	195.99	391.99	783.98	1567.96
G#	207.65	415.3	830.6	1661.2
A	220	440	880	1760
Bb	233.08	466.16	932.32	1864.64
B	247	494	988	1976

7.10 Common tuning chart based on A =440 Hz
and notes based on Equal Temperament.
Source: Sonicgeometry.com

Note Vibrational Frequency of Each Octave

C	126	252	304	1008	2016
C#	135	270	540	1080	2160
D	144	288	576	1152	2304
D# (Eb)	153	306	612	1224	2448
E	162	324	643	1296	2592
F	171	342	684	1368	2736
F#	180	360	720	1440	2880
G	189	378	756	1512	3024
G#	198	396	792	1584	3168
Ab	207	414	828	1656	3312
A	216	432	864	1728	3456
A#	225	450	900	1800	3600
B	234	456	936	1872	3755

7.11 Pythagorean Tuning Based on A = 432
Hz and notes based on 3:2 ratio
Source: Sonicgeometry.com

The three angles of a triangle add up to 180 degrees. Using the Pythagorean tuning system with the number 432 as the basis for tuning to the note A, a sound with a vibrational frequency of 180 hertz, or 180 vibrations per second, produces the note of F# (F sharp). A circle has 360 degrees, and the sum of the angles of a square is 360 (degrees) as well. A vibrational frequency of 360 hertz produces the sound F# once more, only 1 octave higher. 180 + 180 = 360. The angles of a pentagon add up to 540 degrees and 540 hertz produces a C# (C sharp), a harmonic fifth of F#. The degrees of a hexagon's angle total 720 degrees, and 720 vibrations per second results in another F#, again one octave higher than the circle and square and two octaves higher than the triangle. A septagon is equivalent to 900 hertz which is an A#. These geometric shapes in their corresponding vibrational frequencies produce a perfect F# major chord in 3 part harmony. When one adds an octagon to the mix, a vibrational frequency of 1080 hertz produces another C#. This now produces a 3 part major chord in the key of F#.

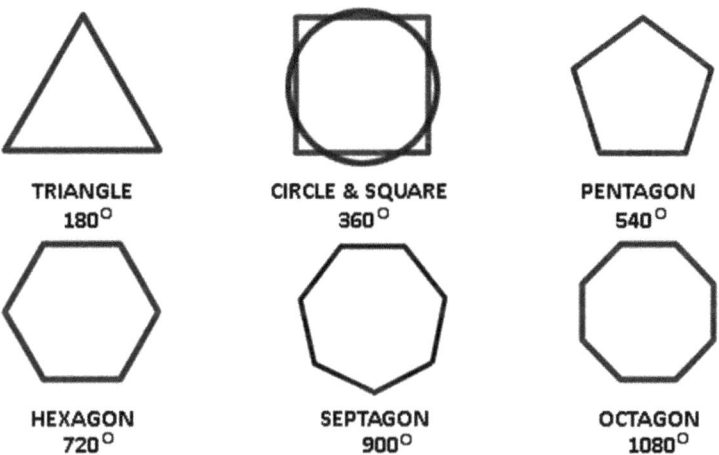

7.12 Triangle, Circle and Square, Pentagon, Hexagon, Septagon, Octagon with corresponding angle sum.

What about three dimensional shapes? A Platonic solid is a regular, convex polyhedron. It is constructed by congruent, regular, polygonal faces with the same number of faces meeting at each vertex. Five solids meet these criteria (Figure 7.13).

A tetrahedron is the simplest platonic solid that uses straight lines. It consists of 4 triangular faces and 12, 60 degree angles, which total 720 degrees. Using Pythagorean tuning, 720 hertz produces an F#. A cube is 2160 degrees of angles which is a C#, an octahedron is 1440 degrees which is another F#. An icosahedron is 3600 degrees, and 3600 hertz is an A#, which completes another F# chord.

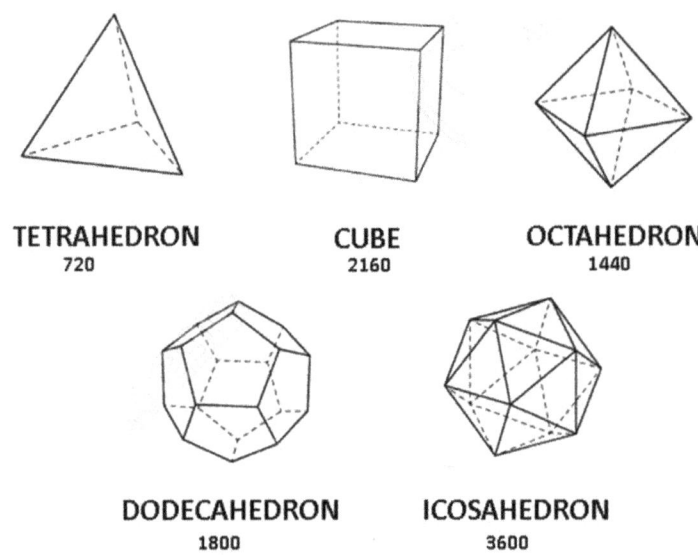

TETRAHEDRON
720

CUBE
2160

OCTAHEDRON
1440

DODECAHEDRON
1800

ICOSAHEDRON
3600

7.13 The 5 Platonic Solids with corresponding angle sum in degrees.

If you had a means of seeing sound, what would it look like? Sound is a vibration, but as de Broglie demonstrated, everything is vibration and what you experience in your everyday life is a manifestation of these vibrations as perceived by your senses. If you were able to see an F sharp it would not look like a triangle or a tetrahedron; it would look like a geometric pattern similar to a mandala (Figure 7.14).

Every object has a characteristic frequency, or frequencies, at which it vibrates most, with the least input of energy. Have you ever noticed how sand or other particles on a vibrating surface seem to form patterns? The vibrating surface forms standing waves called modes from which particles move away. They collect in areas of least vibration, or no vibration, called nodes. The higher the frequency of the vibrations, the more complex are the patterns formed.

7.14 Two examples of a type of Mandala. (There are also religious mandalas and fractal mandalas).

Some may say the correlations shown between geometry and vibrational sounds are a function of a random choice of measurement, that they are simply coincidences that disappear if one chooses another counting system. One should consider, however, that the basis 12 and 60 measurement system was used by the Sumerians at least 6000 years ago and was taught to them by the Annunaki. This system gives us the 360 degrees in a circle and the base number for all geometry. There are 60 minutes in an hour and 60 seconds in a minute. There are 12 inches to a foot, 12 months to a year, 12 signs of the zodiac, 12 Tribes of Israel, 12 apostles in Christianity, 12 Olympian gods and 12 Titans in Greek mythology and the list goes on: there are 12 hours in a day and 12 hours of night. As you can see, our world is built around these numbers. There are also 12 notes in an octave and 12 fundamental particles that make up our physical world. It is obvious that a great deal of symbolism is based on the number 12. But why?

The standard idea for the shape of the universe has been that it is infinite and tending toward flat. In 2003, a group of scientists published a paper entitled: *Dodecahedral space topology as an explanation for weak wide-angle temperature correlations in the cosmic microwave background.* In this paper, the case is made for the universe to be a 12 sided dodecahedron based on the cosmic microwave background, which provides a picture of the universe as it was some 400 000 years after the big bang. Any variations in the temperature of the background radiation reflect variations in the density of the universe at this time. These temperature fluctuations can be expressed as a sum of spherical harmonics, and astrophysicists plot the relative strength of these harmonics as a function of angle. The height and positions of the peaks in this 'power spectrum' are related to the basic astrophysical properties of the universe. A dodecahedron has a total of 1800 degrees which, according to Pythagorean tuning, would be an A#. As theoretical physicist Michio Kaku states, "The mind of God would be cosmic music resonating through hyperspace". Referring back to our Pythagorean grid in figure 7.11, we see the grid numbers and their bases recurring over and over again in our world. However, some of these occurrences do not seem to be natural.

We described some of the surprising numerical occurrences in chapter IV when discussing our Moon. Here are some more based on the factor 432 tuning system.

If one looks at the grid in figure 7.11 once again, one can see that the base number 216 is the note A below middle C and half of the "magical" keynote number 432. The diameter of the Moon is **2,160** miles. Multiply that by 400, and one gets 864,000, which is the diameter of the Sun in miles. Of course 864 is twice 432, so also a note A.

Divide the Moon's diameter by 2: 2160 miles / 2 = 1080, the angle sum of an octagon.
By 3: 2160/3 = 720, the angle sum of a hexagon.
By 4: 2160/4 = 540, the angle sum of a pentagon
By 5: 2160/5 = 432, the keytone of the Pythagorean grid.
By 6: 2160/6 = 360, the number of degrees in a circle and the angle sum of a square.

There are **86,4**00 seconds in a day.

432^2 (432 squared) = 186624, the speed of light within 0.01%.

The Earth's precession of 25920 years divided by the 12 constellations of the zodiac that it passes through in its journey = 2160, the diameter of the Moon in miles (2160 is a C#).

It is apparent that these relationships could not be due to random chance. The interconnection of the geometry and mathematics involved demonstrates a knowledge of the universe and capabilities far beyond those of the human race.

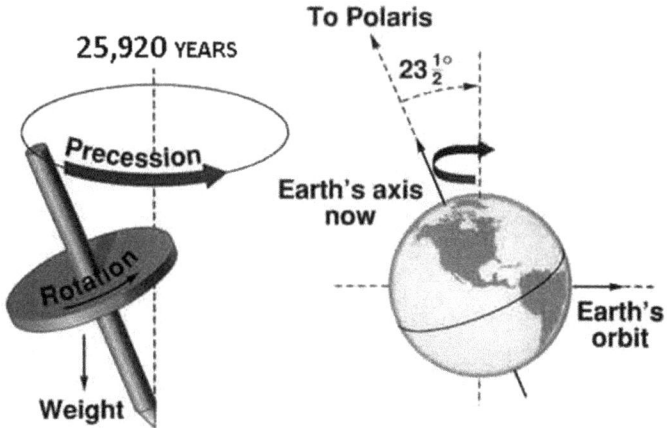

7.15 Earth's Precession. The Earth's Wobble on its axis takes 25920 years to make a complete rotation as it progresses through the 12 constellations of the Zodiac. Diagram after Timothy Connelly.

THE FLOWER OF LIFE

The Flower of Life symbol is another example of a mandala. It is found throughout the world; in Spain, in Italy, the Forbidden City complex of Beijing, China, synagogues in Israel, in Buddhist temples of India and Japan, in the ancient city of Ephesus in Turkey, and in the Temple of Osiris in Abydos, Egypt. (Figure 7.16).

7.16 The Flower of Life Pattern

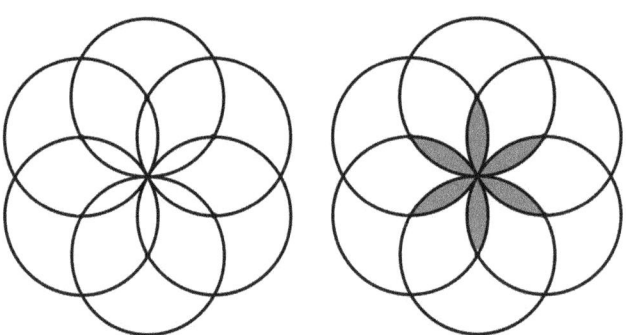

7.17 The Seed of Life

With the Flower of Life, we once again see sacred geometry at work. The pattern begins with the Seed of Life, which consists of six overlapping circles (Figure 7.17). By referring to the Pythagorean grid and assigning a frequency to the degrees, a progression of tones is revealed. The first circle is 360 degrees and is an F# (F sharp); two circles is 720, another F#; 3 circles is 1080, a C#, and a harmonic fifth; 4 circles, 1440, another F#; 5

circles is 1800, an A#, providing the harmonic third of an F# major chord; and 6 circles is 2160, another C#. The center of the six circles produces 6 vesicles, each composed of two 60 degree arcs. The shape of the vesicle also depicts the action of a plucked or vibrating string.

If one adds another arc in the 3rd dimension, the figure created has three equal edges and three equal faces, a figure that some would call, the simplest platonic solid. Artist Michale Evens named this form the Trion Re' (Figure 7.18). It was previously mentioned that the tetrahedron is considered the simplest platonic solid since, in classic mathematics, a platonic solid is a convex polyhedron composed of straight lines and flat faces, with the same number of faces meeting at each vertex. However, since space is curved, there really is no such thing as a straight line or a flat face. Therefore, the Trion Re' could arguably be the simplest platonic solid.

The fact that the Flower of Life pattern is found in many ancient cultures reveals the importance of this design. Could it be telling us that all things are made up of quantum vibrating strings of energy that come together to form geometrical patterns of harmonic structure? This is known as string theory in theoretical phyisics today.

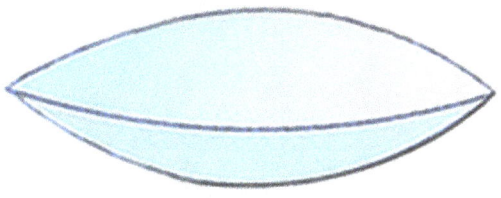

7.18 Trion Re'

It has been known for some time that geometry plays an important role in the structure of many living things. The Golden Spiral is formed by a number sequence whose growth factor is ϕ (phi) (1.61). A golden spiral gets wider (or further from its origin) by a factor of ϕ for every quarter turn it makes (Figure 7.15). This mathematical sequence is found in flower petals, seed heads, shells, tree branches, hurricanes, galaxies, the human uterus, and the DNA molecule, to name only a few (Figure 7.19).

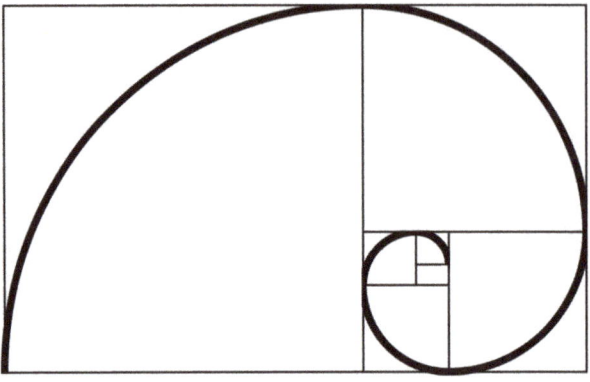

7.19 Golden Spiral

7.20 Examples of the Golden Spiral in nature.
Image courtesy designmantic.com

Another place the swirling structure of the Golden Spiral is found is within the human pineal gland. A small endocrine gland, the pineal gland lies near the center of the brain, and is the only organ, besides the eyes, to contain light receptors. These receptors are likely one of the reasons the pineal gland is called the third eye. A symbolic depiction of the third eye, located in the center of the forehead, is found in religious drawings and sculptures of religions, such as Buddhism and Hinduism (Figure 7.17). The known function of the pineal gland is the production of melatonin, the sleep hormone. This would obviously be one function of the light receptors in the gland. Perhaps there are others of which the ancients knew, and we have yet to rediscover.

7.21 Statue of Lord Shiva with Third Eye
Depicted in the center of the forehead

Looking at the winged deity Marduk in figure 7.22, one can see that he holds in his right hand something that looks like a pine cone. It is believed that this is a representation of the pineal gland, whose name is derived from the Latin *pinea*, the word for pine cone.

On the right, one finds the same object, accompanied by wings and entwined serpents in the form of the Caduceus. Displayed in this one image is the pineal gland, the shape of the DNA molecule, a double helix formed by the twin serpents, and the shape of the vesicle or plucked string, also formed by the entwined serpents. As well, the wings and the two serpents present the bird - serpent paring we have continually encountered.

So, is the symbolism of the square and compasses within the Starman's nose ornament merely emblematic jewelry, or are we being told that geometry is an essential key to unlocking the secrets of the universe?

7.22 Left: Babylonian Winged Deity, Marduk.
Right: Caduceus, ancient symbol still used in medicine today.

FOOTNOTES:

1. J.E. Cirlot, *A Dictionary of Symbols*, translated from Spanish by Jack Sage (London: Routledge, 1971) 380
2. Christopher Powell, "The Shapes of Sacred Space," lecture at 19th Annual Weekend, University of Pennsylvania Museum, March 24, 2001.
3. Richmond Lattimore, The Iliad of Homer (Chicago: The University of Chicago Press, 1951) 180
4. Richard Aldrington and Delano Ames, translation, *The Larousse Encyclopedia of Mythology* (New York: Barnes & Noble, 1994) 115, 116
5. https://en.wikipedia.org/wiki/Louis_de_Broglie
6. https://www.merriam-webster.com/dictionary/geometry
7. https://io9.gizmodo.com/15-uncanny-examples-of-the-golden-ratio-in-nature-5985588
    * https://soundofgoldenlight.com/432-hz/

CHAPTER VIII

THE BEGINNING AND NEARLY THE END

I am Alpha and Omega, the first and the last, the
beginning and the end, – Revelation 22:13

THE CRADLE OF CIVILIZATION

So, what made us forget our past? And what led to the apparent destruction of civilization on the planet Mars? We must go back to the time when the gods first came to Earth in a hope to answer those questions. Using the time frame established by Sumerian texts, the late author and historian Zecharia Sitchin dates this to roughly 445,000 years ago.

The area where this occurred is Mesopotamia, the ancient region of Western Asia. Situated within the northern part of the area known as the Fertile Crescent, it encompassed the areas currently corresponding to Iraq, Kuwait, eastern Syria, southeastern Turkey, and regions along the Iran – Iraq border. This area is considered the cradle of civilization, and in the late 19th century, ancient cuneiform tablets, dating back over five thousand years, were discovered in this region. It is estimated that at least half a million cuneiform tablets have been excavated in modern times,[1] of which only about 10 percent have been read.[2] Before these tablets were discovered, the Bible was considered the oldest of the ancient writings, but the publishing of these Mesopotamian texts transformed our understanding of history.

The translation of the Epic of Gilgamesh, in 1872, by the scholar and translator George Smith, allowed other cuneiform tablets to be interpreted.[3]

This altered the understanding of the biblical version of history existing at the time and provided a new pathway for exploring our past. In the mid-20th century, scholars W.G. Lambert and A.R. Millard translated an ancient Assyrian text whose opening line reads, "When the gods, like men, bore the work". They concluded that the text was based on earlier Sumerian versions and possibly on even earlier oral traditions. It told a story about the arrival of the gods on Earth, the creation of Man, and the attempt to eliminate him via the Deluge. [4]

In the mid to late 20th century, Zecharia Sitchin became one of the first historians to regard these ancient texts as recorded history rather than mythology. He learned the languages, history, and archaeology of the ancient Near East and wrote numerous books, eight of them collectively known as The Earth Chronicles. The books describe an ancient time when the gods came to Earth, had offspring, and warred amongst themselves.

Most ancient civilizations believed that their ancestors descended from the heavens. If one considers the heavens were the cosmos, it is easy to see the gods as space travelers, not mythical beings of human imagination. The stories from ancient texts, to a large extent, are likely true, but due to the limited knowledge of science and technology at the time, the details may seem rather fanciful.

The beings or gods described in the texts and stories were referred to as immortal, at least functionally, and their life spans on Earth certainly attest to that. They had obviously found a way to rejuvenate their DNA, one secret we humans are just beginning to uncover.

The ancient texts of Mesopotamia refer to Anu, the supreme being of the sky, whose abode was heaven. His two sons, born of different mothers, were EA (which means "Whose House Is Water) and Enlil (Lord Wind). Although EA was the oldest, Enlil was born of Anu's paternal half-sister, giving him the necessary royal blood to be the successor to Anu.

EA was the leader of the first group of space travelers to come to Earth, and he set up an Earth station called Eridu. He was thus given the title EN KI, meaning Lord of Earth. A Sumerian text describes EA's (Enki's) arrival:

> *When I approached Earth there was much flooding.*
> *When I approached its green meadows, heaps and mounds*
> *were piled up at my command.*

I built my house in a pure place…
My house- its shade stretches over the Snake Marsh…[5]

When his father, Anu, and half-brother, Enlil arrived, rulership over the land was given to Enlil, while EA (Enki) was assigned dominion over the seas. This did not meet the approval of Enki, and although there may have previously been animosity between the two brothers, giving control of Earth to Enlil was the beginning of many future conflicts between Enki, Enlil, and their descendants. Not the least of these conflicts would involve humans and our very existence.

THE DEITIES OF MARS

The gods were known to the Sumerians as the Anunnaki. When they first came to Earth, a portion of the crew remained in the mother ship while others transported down. From a Sumerian text:

Assigned to Anu, to heed his instructions,
Three hundred in the heavens he stationed as a guard; the
ways of Earth to define from the Heaven;
And on Earth Six hundred he made reside.
After he all their instructions had ordered, to the Anunnaki
of Heaven and of Earth he allotted their assignments.[13]

Those who stayed on the orbiting spacecraft were referred to as the Igigi. In a hymn to the chief of the shuttlecraft crews, Shamash, is a description of his return to the spacecraft:

At thy appearances, all the princes are glad
All the Igigi rejoice over thee….
In the brilliance of thy light, their path….
They constantly look for thy radiance…. Opened wide is the
doorway…. the bread offerings of all the Igigi (await thee).[14]

Camera systems orbiting Mars and limited rover exploration, have showed us that Mars once had rivers and oceans, although we do not know when they disappeared. It would therefore seem likely that when the Anunnaki voyaged into our solar system, they would have first stopped at the planet Mars.

Sitchin believes that a way station was set up on Mars to accommodate the coming and going from Earth to their home planet. It would be the Igigi that would occupy and run the way station, and this may very well have been the place where the gold was refined before shipping. Figure 8.1 is a reproduction of an ancient Sumerian cylinder seal depicting a spacecraft or communication satellite. A fish-man, the symbol of the followers of Enki, is on the right accompanied by a six-pointed star, the symbol for the planet Mars. The planets were counted from the outside towards the Sun, in the order one would encounter them when coming from outside the solar system. Mars was the sixth planet, Earth the seventh, and, according to Sitchin, the seven circles accompanied by the crescent depict the person on the left as being on Earth.

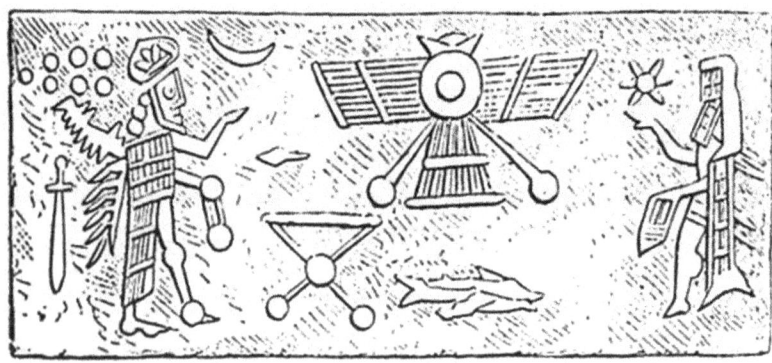

8.1 Image from a Sumerian Cylinder Seal seemingly showing communication between beings on Earth and Mars.

The person depicted on Mars as an Enkiite comes as no surprise as it was EA (Enki) who commanded the first Expedition to Earth, and he likely would have setup the way station. In ancient Sumer, Enki's son, Nergal, was associated with the planet Mars and was in charge of the African domains, which included the gold mining areas in the south. Marduk, Enki's oldest son, was associated with the planet Mars during his reign as the main god of Babylon. Marduk was also very likely, the Egyptian god Ra, who became Amen Ra (Hidden Sun) when he was exiled by Enlil for his warring ways. It is not clear where he was exiled to; was it Mars or, perhaps Mesoamerica. Another important fact regarding Marduk is that his main symbol was a dragon. Furthermore, followers of his uncle, Enlil, referred to him as the Great Serpent.

As stated in chapter VI, the Mayan glyph for Mars is known to archaeologists as the Zip Monster. Also, as shown in figure 6.8, within the Mars glyph are a half glyph of a bird, and a half glyph of a serpent. The bird-serpent pairing was the symbol for the Mesoamerican god Quetzalcoatl (Kukulkan) and it is also found within the Egyptian winged globe insignia (figure 8.2).

8.2 Egyptian Winged Globe Symbol (Bird and Twin Serpent Pairing)

8.3 Left: The winged Babylonian god Marduk holding a vessel in his left hand. Drawing by George J. Haas (image source, Larousse Encyclopedia of Mythology) Right: The winded Maya and Aztec god Quetzalcoatl holding a vessel in his left hand. Drawing by George J. Haas, (image source: Mysteries of the Mexican Pyramids)

Now, compare the images in Figure 8.3 of Marduk and Quetzalcoatl. Depicted are two gods from ancient cultures on opposite sides of the world, both holding vessels in their left hands, both winged, both with epithets of serpents, and both associated with the planet Mars. In his book, *The Lost Realms*, Sitchin proposes that Quetzalcoatl was Ningishzidda, Enki's youngest son, who was given the secrets of life and death along with his father's symbol of twin serpents. In chapter II, I suggest that Lord Sun (the Jaguar god) is also Enki, as he is surrounded by sea creatures, a common depiction of him in Sumerian images. I state that since he is accompanied by two other versions of Quetzalcoatl, the Feathered Serpent, and the Vision Serpent, that he, himself, was Quetzalcoatl. So we have evidence for three separate deities, Enki and two of his sons, all as candidates for the god Quetzalcoatl. Therefore, whether it was Enki himself, Ningishzidda, his youngest son or Marduk, his oldest son, an Enkiite was a main deity of Mesoamerica and of Mars.

THE CREATION OF MAN

The Anunnaki arrival predated human existence on Earth. This group of travelers descended to the planet with plans to colonize and mine its resources, particularly gold.*

It was the hard labour that drove them to create a worker to toil for them. When the work in the mines and elsewhere became too great a burden for the lesser gods, they rebelled, appearing at the very gates of his residence and threatened mutiny against the great god Enlil.

Fearing the worst, a council of the great Anunnaki was assembled, with Anu himself coming down to Earth. It was Enki, who came up with a solution to the problem, and during the gathering stated:

> *"While the Birth Goddess is present*
> *Let her create a Primitive Worker;*
> *Let him carry the toil of the gods!"*[6]

---

* (Gold's value is not seen for its currency, but because of its unique physical and chemical characteristics. It is the most malleable and ductile of all the metals, has the highest corrosion resistance of all the metals and it does not oxidize.)

Problems arise in translating languages when words have more than one meaning or when they are used metaphorically. The Bible states that God made Man from clay, but this is most likely a mistranslation or a play on words. As Sitchin states, "The Akkadian term for clay or molding clay is tit, but its original spelling was TI.IT, meaning that which is with life". So it was not actually clay, but something that was "with life". It was an existing hominid that Enki would "bind upon it the image of the gods". The original biblical text does not use the term God, but rather Elohim, which means deities and was written in the plural form.

> *And the Elohim said: Let us make Man in our image, in our likeness.* [7]
>
> *And the Elohim, fashioned the Adam from the clay of the soil; and he blew in his nostrils the breath of life, and the Adam turned into a living Soul.*[8]

According to Sitchin, "the Hebrew term commonly translated as 'soul' is *nephesh*, that elusive 'spirit' that animates a living creature and seemingly abandons it when it dies."[9]

It was Enki, whose symbol was twin serpents (DNA ?), and his sister Ninhursag, the chief medical officer, who would work together to create Man. There was much experimentation before a satisfactory "Adapa" was created. It would be Enki's wife, Ninki, whose womb would be used. However, according to Sitchin, an embryo produced from the fertilization of a female Homo Erectus egg by the sperm from an Annunaki, likely Enki himself, was probably implanted into Ninki.

Sitchin calculates the creation of the Adapa to be about 300,000 years ago, in the Shimti, "the house where the wind of life is breathed in", which was in Southeast Africa where the mining was ongoing.[10] Not surprisingly, modern mitochondrial DNA science has traced a human 'Eve' back to southern Africa to at least 200,000 years ago.

Sitchin points out that throughout the course of human events described in the Sumerian texts, it is Enki who emerges as Mankind's protagonist and Enlil as its discipliner, if not outright antagonist. It was Enki who created humans in South Africa, and it was Enlil who wanted the workers to tend his garden in Mesopotamia, in EDIN (Eden). Sitchin states, "The role of a deity wishing to keep the new humans sexually suppressed, and of a deity willing and capable of bestowing on Mankind the fruit of 'knowing', describe Enlil and Enki perfectly."[11] He goes on to say:

> "*Once more, Sumerian and biblical plays on words come to our aid. The biblical term for 'Serpent' is nahash, which does mean snake. But the word comes from the root NHSH, which means 'to decipher, to find out'; so that nahash could also mean 'he who can decipher, he who finds things out', an epithet befitting Enki, the chief scientist, the God of Knowledge of the Nephilim.*"[12]

The new workers were in scarce supply, and the Annunaki in Sumer wished to acquire their services. Since they were sterile hybrids, it was this demand that led Enki to give Man the ability to procreate.

He granted them, that which the Bible refers to as the "knowing" of good and evil. In the Old Testament, the term "to know" is used to refer to sexual intercourse between a man, and his wife, for the purpose of having children. It was, therefore, the god Enki (the serpent) who gave Mankind life and the ability to procreate (eat from the forbidden fruit). It would be Enki as well, who would save us from extinction in the great flood.

THE GREAT FLOOD

Did the destruction of Mars occur at the same time as the great flood on Earth, an event that some suggest was at the end of the last glacial period, some eleven or twelve thousand years ago? The biblical story of Noah and the deluge is one that was passed down from earlier civilizations. The same tale is found in the Sumerian Epic of Gilgamesh, where Ziusudra is the hero, and in the Akkadian epic Atra-Hasis, where it is Utnapishtim. As discussed in chapter II, the Mesoamericans had their own version of the flood.

According to the Bible, the great flood came at a time when God became disenchanted with the evils of Man and the genetic deterioration due to breeding with "the sons of God". The exact wording of Genesis, chapter 6 varies with different versions of the Bible, but the essence is the same:

And it came to pass, When the Earthlings began to increase in number upon the face of the Earth, and daughters were born unto them that the sons of the deities saw the daughters of the Earthlings that they were beautiful, and they married any of them they chose.[15]

The Book of Enoch is a Jewish text and part of the Dead Sea Scrolls found in the Qumran caves on the north shore of the Dead Sea in 1947. It is accepted that the text was written between 400 B.C.E. to 300 C.E. In chapter 7 of the text it states:

It happened after the sons of men had multiplied in those days that daughters were born to them, elegant and beautiful. And when the angels, the sons of heaven, beheld them, they became enamoured of them, saying to each other, come, let us select for ourselves wives from the progeny of men, and let us beget children...

Their whole number was two hundred, who descended upon Ardis, which is the top of mount Armon...

In Judeo-Christian tradition, these "sons of the deities would become known as "the fallen angels". In actuality, they were the Igigi, those who were originally on the Mother Ship, and later, stationed on Mars. In *The Lost Book of Enki*, Zecharia Sitchin suggests that life on Mars was becoming difficult. Mars was becoming drier, and the number of females was few. The Igigi wanted to abandon the station on Mars and come to Earth to take human brides; this was not allowed. However, when Marduk married the human woman Sarpanit, a large number of Igigi came from Mars to attend the wedding. Sitchin writes;

In a great number did the Igigi from Lahmu (Mars) to Earth come, only one third of them on Lahmu stayed, to Earth came two hundred. To be with their leader Marduk,

his wedding celebration to attend, was their explanation;
unbeknownst to Enki and Enlil was their secret: to abduct
and have conjugation was their plot.[16]

After the wedding, the Igigi hatched their plan and abducted women of
their choosing.

To the landing place in the Cedar Mountains (Lebanon)
the Igigi with the females went, into a stronghold the place
they made, to the leaders a challenge they issued: Enough of
deprivation and not having offspring! The Adapite daughters
to marry we wish. Your blessing to this you must give, else by
fire all on Earth destroy we will!

Enki and Ninmah their heads shook, with begrudging
agreement they voiced. Only Enlil was enraged without
pacification.[17]

Egyptian hieroglyphs depict the Annunaki as a people of great physical
stature, and it seems that the conjugal pairing of the Igigi and the human
women also produced giant offspring. Known by various terms in the Bible,
such as Nephilim (giants) or Gibborim (mighty), they were eventually
referred to as demons. Referring to the time of the deluge, the Bible states:

The Nephilim were on the earth in those days- and also
afterward – when the sons of God went to the daughters of
men and had children by them.[18]

What could produce giant offspring from the pairing of the Igigi and
humans? If we are to believe that Enki and his staff genetically engineered
humans, then we would expect the process to have been well controlled.
The Sumerian texts do describe some freakish results from some of the
first attempts at creating the *lulu amelu (primitive worker)*. Is it possible
that uncontrolled interbreeding produced unexpected hybrids of even
larger size? Could the Igigis' being on Mars for thousands of years have
affected their DNA? Perhaps Mars, having less gravitational pull than
Earth, affected the genes of the Igigi. Whatever the reason, it is likely
a true occurrence as a reference to the giants is found within all the

ancient writings. Moreover, they became a menace to the gods, and it was Enlil's desire to eliminate the giant hybrids, along with humans, with the flood. The Annunaki perceived a difference between destiny and fate. Destiny you cannot control; it is the predetermined course of events that cannot be changed, such as the motion of the heavens. Fate, however, is a predetermined course of events that can be altered.

To Enlil it was destiny that the great flood would occur. It was something he foresaw and could not change. The annihilation of the Nephilim and humans by the great flood however, would be their fate.

It is hard to imagine any single event that could flood the entire world. Despite the Aztec story of fifty-two years of rain or the biblical story of 40 days and nights of rain, the Earth's hydrologic cycle dictates that the amount of water on the Earth and in the atmosphere remains basically the same. So where would the extra water come from? Perhaps a rapidly melting ice sheet from a solar flare or an asteroid strike. Perhaps it was the sliding of the Earth's crust on the molten mantle causing a massive tsunami. Whatever the cause, if the flood were destiny, it would have had to be an unstoppable event.

Man's capacity to procreate angered Enlil, but the marriage and interbreeding with the Igigi enraged him. Thus it was determined that the giants, as well as humans, were to be destroyed by the deluge. In the pre-biblical flood stories of the Sumerian, Babylonian, and Akkadian texts, it was Enki that alerted the hero, the biblical Noah, to build a ship to withstand the flood and save the seeds of all living things. Enlil would not discover until after the flood had subsided that not all of Mankind had perished. Once again, he was angered but came to realize that if annihilation was not Mankind's fate, then survival must be its destiny.

But how would a flood on Earth destroy civilization on Mars? It obviously wouldn't, but it may have led to another important event; the transfer of humans to Mars. Perhaps the Enkiites saved more humans from the deluge by transferring some of Earth's human inhabitants to Mars. Maybe that is why their religious mythology is found on that planet! If you recall from chapter II, the Aztec goddess, Chalchiuhtlicue, is said to have caused the flood, but she is also said to have created a bridge linking Earth to Heaven for the souls of those she favored. Is this a reference in the understanding at the time of the transport of some of the Mesoamericans to Mars? So then, what caused the destruction of the Mars civilization, and when?

THE PYRAMID WARS

The struggle for control amongst the gods is well documented in Sumerian and Akkadian texts. The 1st pyramid war, as Zecharia Sitchin describes it, was the battle between Horus and Seth. Seth had been given the rulership over Upper Egypt in the south, while his half-brother, Osiris, ruled the northern, Lower Egypt. Seth, however, coveted all of Egypt and took the life of Osiris. This story of revenge is a long one and entails the resurrection of Osiris by his wife, Isis. According to some tales, Isis only obtained Osiris's phallus, which allowed her to impregnate herself and give birth to a son, Horus. Other versions say the "essence" of Osiris was obtained. Once again, this may be a misinterpretation of events due to a lack of scientific knowledge. It was quite likely, the procurement of Osiris' DNA allowed Isis to be impregnated and give birth to his heir. When Horus was older, he went to war with Seth to revenge his father's death and regain control of Egypt. He won the battle and gained dominion of all Egypt. Seth's life was spared, and he was exiled to another land.

When the waters had receded after the great flood, all the lands were divided amongst the Annunaki leaders. While the biblical stories relate events of humans, the Sumerian tales focus on the affairs of the gods. As Sitchin states:

> "*We must bear in mind that at the time the allotment of lands and territories was not among the people, but among the gods; the gods, not the people, were the landlords. A people could only settle a territory allotted to their god and could occupy another's territory only if their god had extended his or her dominion to that territory, by agreement or by force.*"[19]

Before the flood, the space facilities were located in what is now southern Iraq, just northwest of the Persian Gulf. The spaceport was in Enki's ancient city of Sippar, while mission control was slightly southeast in

Nippar, the holy city of Enlil. After the flood, all had to be rebuilt. The new location for the spaceport was in the Sinai Peninsula with mission control in Jerusalem (Figure 7.4). The space facilities were of obvious importance, and anyone who desired to rule the Earth needed to control those facilities.

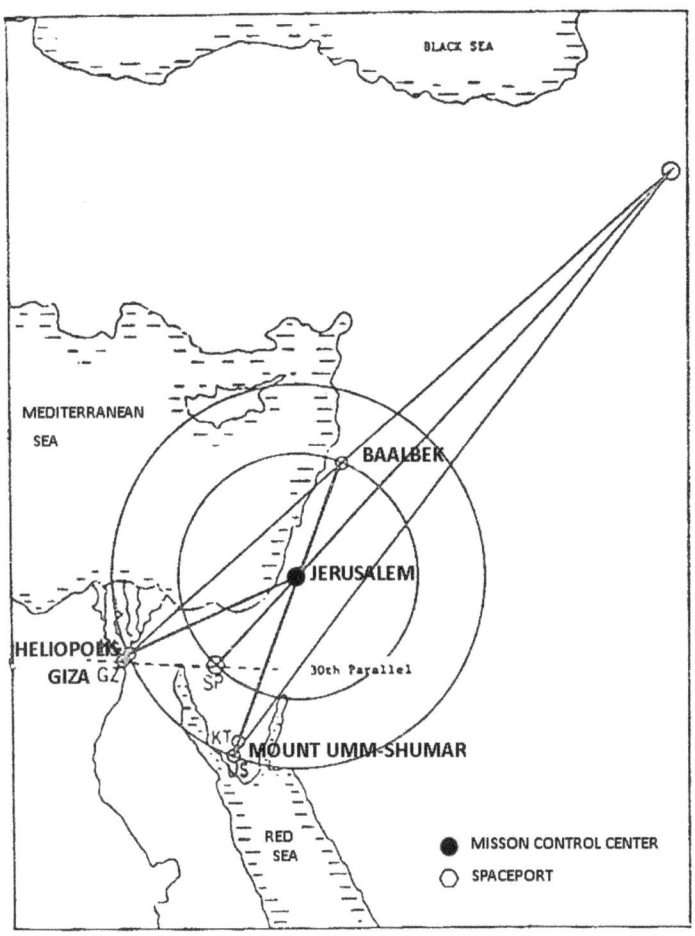

8.4 Location of Spaceport and Mission Control.
(Source; The Cosmic Code, Zecharia Sitchin, p 207)

Seth took up residence in Canaan after he was exiled, which left the space-related areas of Giza, the Sinai Peninsula, and Jerusalem came under Enkiite control. This could not be accepted by Enlilites. The result was the second pyramid war, which Sitchin has dated at about 8670 BCE, 300 years after the first one. A number of writings, including Sumerian, Akkadian, and Assyrian texts, describe the events. The Enilites were led by Enlil's oldest son Ninurta, and the Enkiites by Enki's oldest son Marduk. Battles were fought by human armies overseen by their respective gods. The war was eventually won by the Enlilites, and Enki and his clan retreated to the south and their lands in Africa. Ninurta's conquest was praised in the following Mesopotamian text, collated and edited by Samuel Geller.

> *King, the glory of thy day is lordly;*
> *Ninurta, Foremost, possessor of the Divine Powers,*
> *who into the throes of the Mountainlands stepped forth.*
> *Like a flood which cannot be stopped,*
> *the Enemyland as with a girdle you tightly bound.*
> *Foremost one, who in battle vehemently enters;*
> *Hero, who in his hand the Divine Brilliant Weapon carries;*
> *Lord: the Mountainland you subdued as your creature.*
> *Ninurta, royal son, whose father to him had given might;*
> *Hero: in fear of thee, the city has surrender . . .*
> *O mighty one -*
> *the Great Serpent, the heroic god,*
> *you tore away from all the mountains.*[20]

As we read in this hymn to Ninurta, Marduk is referred to as the Great Serpent, a term also associated with his father, Enki. An examination of the earlier Sumerian stories suggests that the biblical Satan was a consolidation of Enki and Marduk. Although it was his father, Enki, referred to as the Serpent in the biblical Garden of Eden, it was his oldest son, Marduk, who became the usurper of the Throne of Heaven and Earth. His followers were the Igigi, the fallen angels, the demons. The next time Marduk would battle for supremacy would be thousands of years later and would lead to nuclear war.

SODOM AND GOMORRAH

The history and mythology of many of the ancient cultures describe major wars where their gods do battle in the sky. To the followers of the Hindu religion, the gods were unquestionably real; they walked the Earth and co-mingled with humans. The great Hindu god Shiva was both a creator and a destroyer. He created the world and later destroyed it with a great flood. The ancient Hindu texts depict great heroes, demons, and celestial weapons. They describe battles between the righteous and the wicked, and between heroes and demons, in a constant struggle for power.

Discovered in 1947, in a cave above the Dead Sea, by a shepherd boy tending his flock, the Dead Sea scrolls describe a war that is yet to come between the Sons of Light and the Sons of Darkness. This sounds like the narrative found in the Hindu texts. In the Bible, it is spoken of as a war between good and evil or God and Satan, between Israel and its enemies. However, rather than a war yet to come, could this actually have been a war that already transpired? The followers of Judaism, Christianity and Islam are all familiar with the story of Sodom and Gomorrah. One of the most often cited episodes from the Bible, it narrates God's decision to punish the evil of the day by destroying the two cities with fire and brimstone.

If one views the story with modern eyes, however, it becomes quite clear that the text is describing nuclear explosions. But what would incite such drastic action for the gods to use nuclear weapons to destroy these two cities? The Bible tends to define the evil as sexual misdeeds; however, considering the promiscuous and incestuous sex lives of the gods, this seems unlikely. A more plausible explanation is the switching of allegiance by the Kings of Sodom and Gomorrah from Enlil's son Sin (Nannar in Sumerian) to Marduk's son Nabu. The seriousness of the situation became apparent when armies were amassed to march on the space facilities. It seems that Marduk was once again plotting to usurp power on Earth.

A council of the gods was again summoned to assess the situation. Enki was in favor of giving his oldest son Marduk control of the lands and dominion over the gods. However, another of Enki's son's Nergal, was vehemently against it. Although Nergal was a son of Enki, he was married to Ereishkgal, a granddaughter of Enlil, and he had always been envious of his older brother. Nergal, also known as Erra, stormed out of the council, vowing to destroy Marduk. The event was recorded in lines from a Mesopotamia text known as the Erra Epic:

> *"The lands I will destroy, to a dust-heap make them; the cities*
> *I will upheave, to desolation turn them; the mountains I will*
> *flatten, their animals make disappear;*
> *the seas I will agitate, that which teems in them I will decimate;*
> *the people I will make vanish, their souls shall turn to vapour;*
> *none shall be spared . . ."* [21]

From this account it is easy to see why Nergal was known as a god of death, war and destruction.

As the council was at an impasse as to how to proceed, the supreme god Anu was summoned, and the decision to use the seven divine weapons was made.

> *"Those seven, in the mountains they abide,*
> *In a cavity inside the earth they dwell.*
> *From this place with a brilliance they will rush forth,*
> *From Earth to Heaven, clad with terror."* [22]

With this decision the cities of Sodom and Gomorrah, along with the spaceport, would be destroyed. Enlil's son Ninurta and Enki's son Nergal would prepare the weapons. Nergal is quoted once again:

> *"From city to city an emissary (weapon) I will send;*
> *The son, seed of his father, shall not escape;*
> *His mother shall cease her laughter . . .*
> *To the place of the gods, access he shall not have:*
> *The place from where the Great Ones ascend (spaceport)*
> *I shall upheave."* [23]

In the biblical account of the destruction of Sodom and Gomorrah, Abraham stands before God and asks:

> *"Will you sweep away the righteous with the wicked?"* [24]

In the much earlier Mesopotamian text Ninurta speaks with Nergal:

> *"Valiant Erra, will you the righteous destroy with the unrighteous? Will you destroy those who have against you sinned together with those who against you have not sinned?"*[25]

According to Sitchin, a text called CT-xvi-44/46 informs us that a lesser god known in Akkadian as Gibil, whose land adjoined that of Nergal's in Africa, warned Marduk and Nabu of the impending carnage. Marduk and Nabu took heed and fled to the north.

The attack by Ninurta, Nergal and Nergal's attendant Ishum, is described in the Erra Epic.

> *Ishum to Mount Most Supreme set his course; the awesome seven, without parallel, trailed behind him,*
> *At the Mount Most Supreme the hero arrived; he raised his hand – the Mount was smashed.*
> *The plain by the Mount Most Supreme he then obliterated; in its forests not a tree stem was left standing.*
> *Then, emulating Ishum, Erra the King's Highway followed.*
> *The cities he finished off, to desolation he overturned them.*
> *In the mountains he caused starvation, their animals he made perish.*
> *He dug through the sea, its wholeness he divided. That which lives in it, even the crocodiles he made wither.*
> *As with fire he scorched the animals, banned its grains to become as dust.*[26]

With this assault, the space facilities, along with the cities of Sodom and Gomorrah were completely destroyed. Was this a nuclear attack? The manner and description of this holocaust certainly leads one to think so.

In the biblical story, Abraham's nephew, Lot, and his family were spared. They were told to flee and to not look back. Lot's wife did look back, and she was turned into a pillar of salt. This is likely another mistranslation of Sumerian to Hebrew. Salt was obtained from the evaporation of the saltwater in swamps near the Persian Gulf and, in Sumerian, the word, nimur, came to be the term for salt and vapor. [27] It is clear that Lot's wife was vaporized not turned into a pillar of salt.

The Mesopotamian Lamentation texts provide more evidence for the nuclear carnage:

> *On the Land fell a calamity, one unknown to man;*
> *One that had never been seen before...*
> *Causing cities to be desolated, houses to become desolate;*
> *Causing stalls to be desolate.*
> *The sheepfolds to be emptied;*
> *That Sumer's oxen no longer stand in their stalls,*
> *that its sheep no longer roam in its sheepfolds;*
> *That its rivers flow with water that is bitter, that its cultivate*
> *fields grow weeds, that its steppes grow withering plants . . .*
> *The people, terrified, could hardly breathe; the Evil Wind*
> *clutched them, does not grant them another day . . .*
> *Mouths were drenched in blood, heads wallowed in blood . . .*
> *The face was made pale by the Evil Wind.*[28]

The date for this destruction is determined to have been 2024 BCE. Greek and Roman historians describe the valley as having been inundated by the event and even today some water springs have high levels of radiation.[29] Satellite photos of the Sinai Peninsula reveal a massive scar and a surface-deep blackening of limestone rocks.

So, was this the event that destroyed civilization on Mars? There is little evidence to support this as I could find no mention of the planet Mars made in any of the sources. However, there are lines from the Lamentations that state the following:

> *On that day*
> *When heaven was crushed and the Earth was smitten, its face*
> *obliterated by the maelstrom-*
> *When the skies were darkened*
> *and covered as with a shadow . . .*[30]

This statement definitely speaks of two different places. Note, as well, that there is a difference between being crushed and being smitten. Earth was smitten; it was struck with a severe blow, but heaven was crushed! The stars and planets were referred to as the heavens. Is "heaven", in this case, meant to be understood as a single planet...Mars? Marduk was one of Enki's sons who was associated with Mars, and his goal was to usurp power by capturing the space facilities in the Sinai. These facilities suffered the same fate as the rest of the plain. One could assume, therefore, that the space facilities on Mars, already under Marduk's control or at least the Enkiites' control, were also destroyed.

FOOTNOTES:

1. *"Cuneiform Tablets: Who's Got What?"*, *Biblical Archaeology Review*, 2005, https://www.baslibrary.org/ biblical-archaeology-review/31/2/10

2. Lee Watkins, Dean Snyder, (2003), *The Digital Hammurabi Project* (PDF), The Johns Hopkins University, archived (PDF) from the original on July 14, 2014, https://web.archive.org/ web/20140714163229/http://pages.jh.edu/~dighamm/version2/ research/2003_03_Museums%20and%20the%20Web%20 Conference%202003.pdf

3. https://www.britannica.com/biography/ George-Smith-British-Assyriologist

4. Zecharia Sitchin, The 12th Planet (New York: Avon, 1976) 331

5. Ibid. 291

6. Ibid. 334

7. Hebrew Bible, Genesis: 1, 26

8. Ibid: 2,7

9. Zecharia Sitchin, The 12th Planet (New York: Avon, 1976) 349

10. Ibid. 350

11. Ibid. 371

12. Ibid.

13. Ibid. 327

14. Ibid.

15. Hebrew Bible, Genesis: 6,1

16. Zecharia Sitchin, The Lost Book of Enki (Bear and Company, 2002) 200

17. Ibid. 201

18. Hebrew Bible, Genesis: 6,4

19. Zecharia Sitchin, The Wars of Gods and Men (Avon, New York, 1985) 157

20. Ibid. 159

21. Sitchin points out that "…in Ugaritic texts, such as he Canaanite tale of Aqhat (with its many similarities to the tales of Abraham) the death of a mortal by the hand of a god was described as the 'escape of his soul as vapor, like smoke from his nostrils." Zecharia Sitchin, The Wars of Gods and Men (Avon, New York, 1985) 314, 326

22. Ibid. 326

23. Ibid. 328

24. Bible, New International Version, Genesis: 18,23

25. Zecharia Sitchin, The Wars of Gods and Men (Avon, New York, 1985) 327

26. Ibid. 328, 329

27. Ibid. 313

28. Ibid. 336, 337

29. Alan F. Alford, Gods of The New Millennium (Hodder and Stoughton, London, 1998) 326

30. Zecharia Sitchin, The Wars of Gods and Men (Avon, New York, 1985) 338

CHAPTER IX

WHENCE THEY CAME

Can you bind the beautiful Pleiades?
Can you loose the cords of Orion? – Job 38:31

MARS

Beginning in the 19th century, many people believed Mars could hold life. This is due to Italian astronomer Giovanni Schiaparelli's 1877 telescope observation of mysterious straight lines along its equatorial regions, which he called *canali*. The belief persisted mainly due to a mistranslation into English of the Italian canali, meaning channel, and did not necessarily mean a canal or something of unnatural composition.

Up until the late 20th century, popular belief had shifted to Mars as a dead planet that doesn't and never did have life. However, since 1976, orbiting spacecraft and landers have told a different story. At one time Mars had an atmosphere, oceans, and rivers, not unlike Earth.

So, with evidence in this and other books of an advanced civilization having built structures on Earth, Mars, and the Moon, it raises the question; who are/were they, and from where did they come?

It is hard to imagine a civilization more advanced than present day humans evolving on the planet Mars. Being further from the Sun would seem to offer a less hospitable place for life to take hold. There is also the asteroid belt right next door to take into account. One would expect more frequent asteroid strikes on Mars, thus setting back the evolutionary process more often. There is also the subject matter of the geoglyphs or monuments found on Mars. These all relate to earthly animals and mythologies. No, in my opinion, I suspect the travelers to have come from elsewhere.

SIRIUS

The star map and accompanying imagery in chapter V seem to focus on the star Sirius. Is the symbolism of it being a mother star telling us that this is where the builders originated?

More imagery from the bifurcated 'Face on Mars' also points to Sirius. When the humanoid side of the Face from figure 1.4 in chapter I is turned upside down, and the contrast inverted, one is presented with a depiction of the heliacal rising of the star Sirius (Figure 9.2). The heliacal rising of a star or constellation occurs when it rises on the horizon at dawn and is followed by the rising of the Sun. The glyphs on the inverted humanoid side of the face match glyphs from ancient Egypt. At the top is a dog with a semi-circle of dots representing a rising star on the horizon. This obviously depicts Sirius, the Dog Star. Below that is a slightly curved bar shape incised into the surface, representing the 'vault of the sky'. Next is the Sacred Falcon, Horus, leading the Sun, Ra, from the Underworld towards the horizon with the goddess Isis below, arms raised in praise. The 'tear drop' on the cheek of the Maya First Lord becomes the hands of Isis.

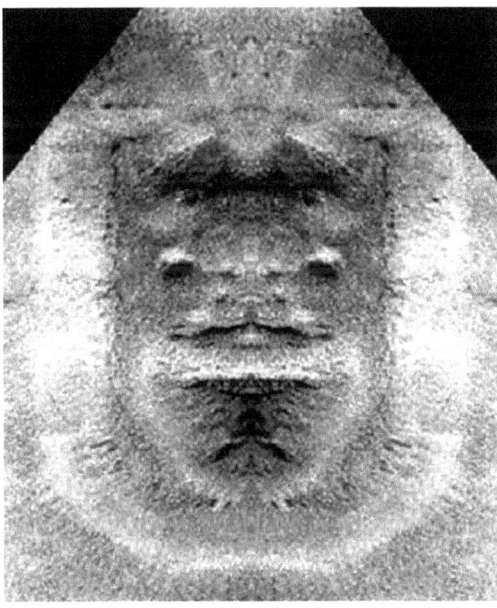

9.1 Mirrored Humanoid side of the Face on Mars from figure 1.4 in chapter I.

187

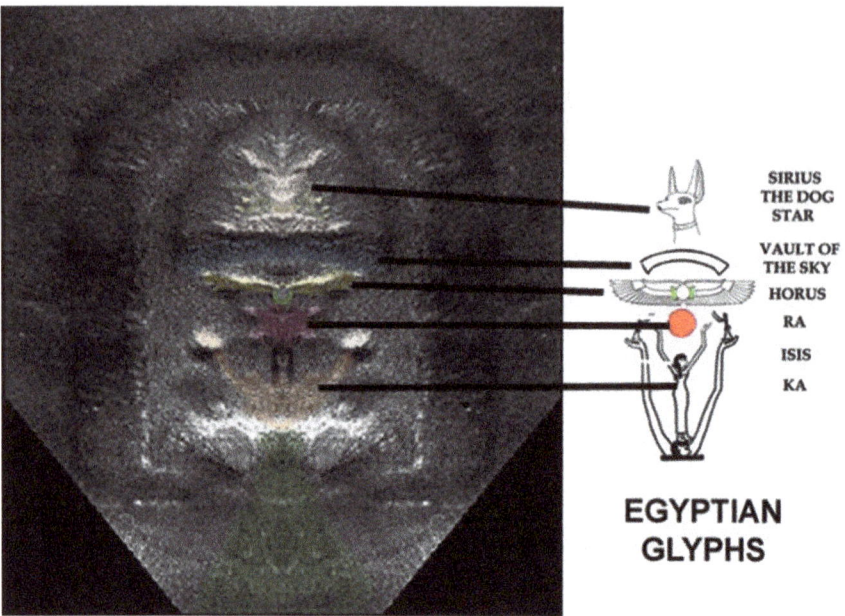

9.2 Heliacal rising of Sirius. Figure 9.1 turned upside down and contrast inverted. False color added by author.

In chapter VI, we mentioned the ancient archaeological site of Gobekli Tepe in Turkey, dating back to over 11,500 years ago. It appears to some researchers that a Gobekli Tepe temple was built to worship or at least follow the star Sirius. Three of the excavated rings are aligned with the points on the horizon where Sirius would have risen in 9100 B.C.E., 8750 B.C.E., and 8300 B.C.E., respectively.[1]

The island of Malta off the south coast of Italy hosts, what are considered to be, the oldest megalithic structures in Europe. These Maltese structures are temples that demonstrate extraordinary craftsmanship. In her book, *Sirius, the Star of the Maltese Temples,* author Lenie Reedijk provides compelling evidence that the temples of Malta go back to 8000 or 9000 B.C.E. and are built to align with the star Sirius.

It seems that reverence for the star was continued in Ireland roughly 3000 to 4000 years later with the construction of Newgrange. In their ground breaking work, *The Island of the Setting Sun*, authors Anthony Murphy and Richard Moore reveal that Sirius was aligned with the winter solstice at the beginning of the temple's construction in 3200 B.C.

In the 1930's two anthropologists, Marcel Griaule and Germain Dieterlen, documented the stories of four priests from the Dogon tribe of Mali, West Africa. The priests seemed to have knowledge that the Star Sirius was accompanied by another star they referred to as Po Tolo and associated it with a tiny grain seed they call Fonio (the botanical name is Digitaria).² Sirius B is a small, dense, white dwarf star, the smallest type of star there is, and because of its size and its distance of 8.6 light years from Earth, it cannot be seen with the naked eye. It was first theorized in 1844 by French astronomer Fredrick Bessel and first observed in 1862 by American astronomer Alvan Clark. So how did the Dogon know the existence of a star that cannot be seen without the aid of a powerful telescope and also associate the smallest type of star known with the smallest seed of which they knew? The priests stated that travellers from Po Tolo, whom they called the Nommo, gave them this knowledge. The tiny Fonio seed is the first of eight seeds given to them by the creator god Amma (Amen), and they were told the star Po Tolo is identical with this seed. According to the Dogon, the Nommo were amphibious and would pass along knowledge to them during the day and go into the ocean at night.

In an area of Sierra Leon, over a 1000 kilometers from the Dogon, numerous small statues of human-like and human-serpent like creatures were excavated during a search for diamonds. Called Nomoli, they were said to represent beings that were sent from Heaven to Earth to live as humans. Due to the depth the statuettes were found, it is determined they could be as much as 17,000 years old. Their construction suggests a high melting temperature would had to have been used in their creation.³ There is also the presence of perfectly spherical steel and chromium balls inside some of the figurines. This suggests the influence of a civilization more highly advanced than humans were believed to be at that time. Chromium was not isolated until 1797.

It seems that an advanced civilization existed on Earth much earlier than historians previously believed and the star Sirius was held in great esteem for thousands of years. So are we to assume that Sirius is the home star system of the extraterrestrials? Our Sun is about 4.5 billion years old and the Sirius star system is believed to be merely 200 million to 300 million years old. So how could life have reached such an advanced stage in such a short period of time?

THE PLEIADES

There are other clues to where the space travellers could have come from. Going back to England, just 2 kilometers from Newgrange another Megalithic site, Dowth Passage Tomb is said to have a curb stone aligned to the heliacal rising of the Pleiades. The Pleiades is a star system in the constellation of Taurus, 444 light years from Earth. In Greek mythology, the brightest stars of the cluster are named for the daughters of Atlas and Pleione. One of those daughters was Maia, which was the original spelling of the current day Maya people. Much like Sirius, this star system gleamed much attention from the ancients. Many Greek temples were oriented to the rising and setting of the Pleiades. The ancient Maya believed their ancestors came from the Pleiades, as do the Hawaiians and the indigenous Cree Nation of Canada. The star system was also known to most all of the ancient civilizations, including the Chinese, Hindus, Native Americans, the Maya, Aztecs, the Maori, the Australian Aborigines and the Japanese who call it Subaru.

The west face of the Pyramid of the Sun at Teotihuacan, Mexico is aligned with the point on the horizon where the Pleiades sets on October 31, the same night where it reaches its highest point at midnight. The Maya Pyramid of Kukulkan in Chichen Itza also has an alignment with the Pleiades. Sixty days after the equinox serpent appears on the pyramid in the spring, both the Sun and the Pleiades reach their zenith above the Pyramid at mid-day.[4]

The Pleiades are mentioned in the Bible and the Hebrew Talmud. In the 19th century astronomer Johann Heinrich von Mädler proposed the Central Sun Hypothesis, according to which all stars revolve around the star Alcyone, in the Pleiades. Based on this hypothesis, until the 1950's, the Jehovah's Witnesses religion taught that Alcyone was likely to be the site of the throne of God. [5]

In the Hawaiian religion, Makahiki is the time to relax, feast and honour the god Lono. It begins when the Pleiades first rises above the horizon after sunset. According, the legends of the indigenous Cree nation in Canada, the Pleiades is from where their ancestors came. Wilfred Buck of the Opaskwayak Cree Nation, and science specialist with the Manitoba First Nations Education Resource Centre in Winnipeg, says their ancestors came to Earth via an umbilical cord from the Pleiades. Buck found many similarities between stories from Cree, Ojibway and Lakota nations and how the peoples saw the sky and their connection to it. "We originate from the stars, we are star people." Buck says:

> *"The genesis mythologies say this is where we come from. We come from those stars, we are related to those stars. Once we finish doing what we come here to do, we go back up to those stars."*[6]

What I find very interesting is he uses the term umbilical cord to describe how their ancestors traveled here from the Pleiades. This seems rather similar to the idea of a worm hole. First proposed in 1916 by Austrian physicist Ludwig Flamm, and later elaborated on by Albert Einstein and Nathan Rosen, a worm hole connects two different points in space-time, theoretically creating a shortcut that could reduce travel time and distance. The shortcuts came to be called Einstein-Rosen bridges, or wormholes.[7] They could connect two separate points in space-time which may be a billion light years apart or a few meters apart. They also could connect two different universes or different points in time. So although the star systems of Sirius, Orion and the Pleiades are vast distances away, perhaps a very advanced raced of beings have learned to harness this theoretical phenomenon.

We saw, in the chapter III star map, the depiction of the star Mirzam acting as ear flares or ear spools on the Sirius Sun-lion. We also mentioned that, to the Maya, those ear flares represented the Sun, symbolized portals to the Otherworld, and were conduits for spiritual energy. Is it possible that stars can act as portals or worm holes?

In trying to determine points of origin, it is noteworthy that the star Sirius, the three stars of Orion's belt and the star cluster of the Pleiades are all in a straight line when viewed on a two dimension star map. Perhaps Sirius and the Orion belt stars are being used to point us to the Pleiades.

As with the Sirius system, we find a very short period of time for development of advanced civilization with the Pleiades. The Pleiades system is estimated to be 75 million to 150 million years old. Perhaps though, life evolves much faster than we think. It did not take hominids

4.5 billion years to evolve on Earth. There have been numerous times where life has been decimated on this planet. It is estimated there have been six major extinction events including 250 million years ago where the Permian – Triassic extinction event eliminated around 96% of all species. As little as 66 million years ago, 76% of all species were wiped out.[9] Perhaps without extinction events on a given planet, it could evolve intelligent life in a relatively short period of time.

NIBIRU

Of course, this discussion of the origins of the Annunaki must include the theory proposed by Zecharia Sitchin. Sitchin believed that the extraterrestrials who created civilization on Earth came from a rogue planet within our solar system. Sitchin maintained that the planet, called Nibiru, orbits our Sun in an extremely large elliptical orbit which takes 3600 years to complete. This orbit takes the planet to a position between Mars and Jupiter during its perihelion and approximately 30 billion miles from the Sun at its aphelion. Sitchin makes some compelling arguments for his hypothesis. However, I find the idea that a human-like species could evolve on a planet with such extreme variations in exposure to the life-giving rays of the Sun very unlikely. I do, however, believe that such a celestial body could exist in our solar system and that it could have, and may still have, a large determination on how our solar system and life on Earth developed.

Despite the discoveries on Earth, Mars and the Moon presented in this book going a long way to corroborating the Sumerian texts, we still do not know for certain where the Annunaki originated. Was it another star system, another galaxy or perhaps even another dimension? The other question is, where did they go? I can't answer that either. However, if they do indeed travel through portals in time and space, then they likely come and go as they please. It is also very likely they have bases on Earth, in or under the oceans.

There are documented accounts throughout history that refer to flying craft in the skies. We know these today as UFOs or UAP (unidentified aerial phenomena). It is possible these craft are of different, visiting galactic species, including the Annunaki, who never left.

FOOTNOTES:

1. https://www.newscientist.com/article/mg21929303-400-world s-oldest-temple-built-to-worship-the-dog-star/
2. Robert Temple, *The Sirius Mystery* (Vermont, Destiny, 1998), 42.
3. https://www.ancient-origins.net/artifacts-other-artifacts/unknow n-origins-mysterious-nomoli-figures-002513
4. https://keysofenoch.org/the-pleiades/ (see also, Maya Cosmogenesis 2012 by John Major Jenkins)
5. http://www.quotes-watchtower.co.uk/god_s_throne_-_pleiades.html
6. https://www.cbc.ca/radio/unreserved/from-star-wars-t o-stargazing-1.3402216/cree-mythology-written-in-th e-stars-1.3402227
7. https://www.space.com/20881-wormholes.html
8. https://cosmosmagazine.com/palaeontology/big-five-extinctions; https://en.wikipedia.org/wiki/Extinction_event

ABOUT THE AUTHOR

William Saunders was born in Edmonton, Alberta, Canada and graduated from the University of Alberta in 1977 with a Bachelor of Science degree. He began working as a geological petroleum technologist in Calgary, Alberta in 1978. He was responsible for petroleum exploration and production for a number of firms over his 35 year career before retiring in 2014. He began his online Mars exploration in 1998 when the new images of the Cydonia area of Mars were being released to the public. He has co-authored two books about the artificial structures on Mars as well as co-authoring numerous papers.

OTHER BOOKS BY
WILLIAM R SAUNDERS

The Cydonia Codex; Reflextions from Mars by George J. Haas and William R Saunders, 2005

The Martian Codex; More Reflections from Mars by George J. Haas and William R Saunders, 2009

DVD Production: The Mars Codex: Manuscript of An Ancient Text, 2014

Papers published

2017

Three-Sided Pyramidal Formation in the Western Region of Candor Chasma. Journal of Space Exploration, Volume 6, issue 3, December 2017. By William R. Saunders, George J. Haas, James S. Miller, Michael A. Dale.

2016

Bearded Profile with Avian Headdress within the Southeast-Facing Slope of an Impact Crater in the Utopia Planitia Region of Mars.

Journal of Space Exploration, Volume 5, Issue 3, November, 2016. By William R. Saunders, George J. Haas, James S. Miller, Michael A. Dale.

A Wedge and Dome Formation Set within the Flat Plains of Libya Montes.

Journal of Space Exploration, Volume 5, Issue 3, November, 2016. By William R. Saunders, George J. Haas, James S. Miller, Michael A. Dale.

Link: HYPERLINK "http://www.tsijournals.com/articles/a-wedge-an d-dome-formation-set-within-the-flat-plains-of-libya-montes.pdf"JSE, Fall, 2016 (Vol. 5 No. 3)*

2015

A Composite Band of Facial Features within a Winding Valley of Libya Montes on the Planet Mars.

Journal of Space Exploration, Volume 4, Issue 2, October 14, 2015, 147-158. By William R. Saunders, George J. Haas, James S. Miller, Michael A. Dale.

2014

Analysis of the Mars Global Surveyor and Mars Reconnaissance Orbiter Images of the Syria Planum Profile Face on Planet Mars.

Journal of Space Exploration, Volume 3, Issue 3, December 30, 2014, 213-230. By J.P. Levasseur, George J. Haas, William R. Saunders, Dr. Horace Crater.

2011

Avian Formation on a South-Facing Slope Along the Northwest Rim of the Argyre Basin.

Journal of Scientific Exploration, Fall, 2011 (Vol. 25, No.3), 515-538.

By William R. Saunders, George J. Haas, Michael A. Dale, James S. Miller, A. J. Cole Dvm, Joseph M. Friedlander Dvm, Susan Orosz Dvm.

INDEX

REFERENCES

"Cuneiform Tablets: Who's Got What?", Biblical Archaeology Review, 2005, https://www.baslibrary.org/biblical-archaeology-review/31/2/10

Alan F. Alford, Alan F., Gods of The New Millennium (Hodder and Stoughton, London, 1998).

Aldrington, Richard, and Delano Ames, translation, *The Larousse Encyclopedia of Mythology* (New York: Barnes & Noble, 1994).

Ancient History Encyclopedia, https://www.ancient.eu/article/415/the-mayan-pantheon-the-many-gods-of-the-maya/

Babayan, Kathryn, *Mystics, Monarchs, and Messiahs: Cultural Landscapes of Early Modern Iran*, (Harvard College, 2002).

Brittin, W. E. (1945). "Valence Angle of the Tetrahedral Carbon Atom". J. Chem. Educ. 22 (3): 145. Bibcode:1945JChEd..22..145B. doi:10.1021/ed022p145.

Campbell, Joseph with Bill Moyers, *The Power of Myth*, (New York, First Anchor Books Edition, 1991).

Campbell, Joseph, *The Mythic Image* (New York: MJF, 1974).

Chevalicr, Jean, and Alain Gheerbrant, *A Dictionary of Symbols* (New York: Penguin Books, 1996).

Cirlot, J.E., A *Dictionary of Symbols* (New York: Barns & Noble, 1995).

Cirlot, J.E., *A Dictionary of Symbols*, translated from Spanish by Jack Sage (London: Routledge, 1971).

de Sahagún, Bernardino, Florentine Codex: *General History of the Things of New Spain, Book 6: Rhetoric and Moral Philosophy.* (School of American Research, 1970).

Freidel, David, Linda Schele, and Joy Parker, *Maya Cosmos; Three Thousand Years on the Shaman's Path* (New York: Quill, 1993).

Gilbert, Adrian D and Maurice M. Cotterell, *The Mayan Prophecies: unlocking the secrets of a lost civilization* (Rockport: Element, 1995).

Gillette, Douglas, *The Shaman's Secret; The Lost Resurrection Teachings of the Ancient Maya* (New York: Bantam, 1997).

Goetz, Delia, and Sylvanus G. Morley (English version) from the translation of Adrian Recinos, *Popol Vuh; The Sacred Book of the Ancient Quiche Maya* (Norman, University of Oklahoma Press, 1950).

Haas, George J. and William R. Saunders, *The Cydonia Codex; Reflections from Mars,* (Frog, Ltd, North Atlantic Books, 2005)

Hebrew *Bible*

Hoagland, Richard C., *The Monuments of Mars: A City on the Edge of Forever,* 5th ed. (Berkeley: North Atlantic, 1992).

http://www.quotes-watchtower.co.uk/god_s_throne_-_pleiades.html

http://www.standrew518.co.uk/Address/View_all.php#bookmark

https://cosmosmagazine.com/palaeontology/big-five-extinctions;

https://en.wikipedia.org/wiki/Chalchiuhtlicue#cite_ref-11 (Berlo 1992: 138; Pasztory 1997).

https://en.wikipedia.org/wiki/Cuban_underwater_city

https://en.wikipedia.org/wiki/Extinction_event

https://en.wikipedia.org/wiki/Eye_of_Ra

https://en.wikipedia.org/wiki/Louis_de_Broglie

https://io9.gizmodo.com/15-uncanny-examples-of-the-golden-rati
o-in-nature-5985588

https://keysofenoch.org/the-pleiades/ (see also, Maya Cosmogenesis 2012 by John Major Jenkins)

https://mayansandtikal.com/mayan-society/mayan-animals-animals/

https://onetribe.net/blogs/content/78343681-status-symbolism-and-spirit -of-the-mayan-ear-flare.

https://soundofgoldenlight.com/432-hz/

https://web.archive.org/web/20140714163229/http://pages. jh.edu/~dighamm/version2/research/2003_03_Museums%20 and%20the%20Web%20Conference%202003.pdf

https://www.ancient-origins.net/artifacts-other-artifacts/unknown-origin s-mysterious-nomoli-figures-002513

https://www.britannica.com/biography/ George-Smith-British-Assyriologist

https://www.britannica.com/topic/Lamashtu

https://www.britannica.com/topic-browse/Animals/Birds/ Eagle-and-Hawk-Order

https://www.cbc.ca/radio/unreserved/from-star-wars-t o-stargazing-1.3402216/cree-mythology-written-in-th e-stars-1.3402227

https://www.merriam-webster.com/dictionary/geometry

https://www.newscientist.com/article/mg21929303-400-worlds-oldes t-temple-built-to-worship-the-dog-star

https://sonicgeometry.com

https://www.space.com/20881-wormholes.htmlhttps://www. thoughtco.com/maya-gods-and-goddesses-117947https://www. transcendenceworks.com/moon-goddess/

https://vesica.org

Kline, Ecuador.

Lattimore, Richmond, The Iliad of Homer (Chicago: The University of Chicago Press, 1951).

Leonard, Jonathan Norton, and the Editors of Time-Life Books, Great Ages of Man; A History of the World's Cultures: Ancient America (New York: Time Inc., 1967).

Masson, Marilyn A., and David A. Friedel, *Ancient Maya Political Economies*, (Walnut Creek, Calf., Altamira, 2002) 63. Also;

Miller, Mary Ellen, *Maya Art and Architecture* (New York: Thames & Hudson, 1999).

Narby, Jeremy, *The Cosmic Serpent: DNA and the Origins of Knowledge* (New York: Tarcher/Putnam, 1999).

Porter III, Frank W, *The Maya* (New York: Chelsea House, 1991).

Powell, Christopher, "The Shapes of Sacred Space," lecture at 19th Annual Weekend, University of Pennsylvania Museum, March 24, 2001.

Richard C. Hoagland, *The Monuments of Mars* (Berkley, Calf. North Altlantic Books, 1992).

Robert Temple, Robert, *The Sirius Myst*ery (Rochester, Vt. Destiny Books, 1998).

Schele, Linda and David Freidel, *A Forest of Kings: The Untold Story of the Ancient Maya* (New York: Quill, 1990).

Schele, Linda and Peter Mathews, *The Code of Kings: The Language of Seven Sacred Maya Temples and Tombs* (New York: Touchstone, 1999).

Schele, Linda, and Mary Ellen Miller, *The Blood of Kings: Dynasty and Ritual in Maya Art* (New York: George Braziller, Inc., 1985).

Sitchin, Zecharia, The 12th Planet (New York: Avon, 1976).

Sitchin, Zecharia, *The Cosmic Code*, Book VI of The Earth Chronicles, (New York, Avon, 1998).

Sitchin, Zecharia, The Lost Book of Enki (Bear and Company, 2002).

Sitchin, Zecharia, The Wars of Gods and Men (Avon, New York, 1985).

Taube, Karl, *Aztec and Maya Myths* (University of Texas Press, 1993).

Temple, Robert, *The Sirius Mystery* (Vermont, Destiny, 1998).

Tompkins, Peter *Mysteries of the Mexican Pyramids; Dimensional analysis on original drawing by Hugh Herleston, Jr. And Historic Illustrations from many Sources* (New York: Perennial Library, 1976).

Vail and Hernandez, *Rain and Fertility Rituals in Postclassic Yucatan Featuring Chaakand and Chak Chel". The Ancient Maya of Mexico: Reinterpreting the Past of the Northern Maya Lowlands.*

Watkins, Lee, Dean Snyder, (2003), *The Digital Hammurabi Project* (PDF), The Johns Hopkins University, archived (PDF) from the original on July 14, 2014.

CPSIA information can be obtained
at www.ICGtesting.com
Printed in the USA
LVHW070247060721
691964LV00025B/411